T0212170

# A Theory of Causation in the Social and Biological Sciences

# A Theory of Causation in the Social and Biological Sciences

Alexander Reutlinger

First published 2013 by
PALGRAVE MACMILLAN

Palgrave Macmillan in the UK is an imprint of Macmillan Publishers Limited, registered in England, company number 785998, of Houndmills, Basingstoke, Hampshire RG21 6XS.

Palgrave Macmillan in the US is a division of St Martin's Press LLC, 175 Fifth Avenue, New York, NY 10010.

Palgrave Macmillan is the global academic imprint of the above companies and has companies and representatives throughout the world.

Palgrave® and Macmillan® are registered trademarks in the United States, the United Kingdom, Europe and other countries.

ISBN 978-1-349-44799-2     ISBN 978-1-137-28104-3 (eBook)
DOI 10.1057/9781137281043

This book is printed on paper suitable for recycling and made from fully managed and sustained forest sources. Logging, pulping and manufacturing processes are expected to conform to the environmental regulations of the country of origin.

A catalogue record for this book is available from the British Library.

A catalog record for this book is available from the Library of Congress.

Transferred to Digital Printing in 2013

*For my parents and my sister*

'Causation is dead. Long live causation!'

Markus Schrenk

# Contents

vii

# List of Figures

# Acknowledgements

Although my name is on the front page, this book has most definitely been collaborative work. This book is a revised version (concerning Chapters 4 and 6) of my dissertation 'The Trouble with Interventions' (University of Cologne, 2011) and, had I not had the opportunity to discuss earlier drafts with the international community of philosophers of science, it would not be in its present form. I will probably not be able to mention all the people that helped me to improve the manuscript. My apologies, and my grateful thanks.

First, I would like to thank my supervisor Andreas Hüttemann for his support and encouragement, and his passion for philosophy. I have the pleasure to be a member of the DFG research group 'Causation and Explanation'. I enjoyed the stimulating company of my friends and colleagues in Köln: Marius Backmann, Siegfried Jaag, Marie Kaiser and Markus Schrenk. It has been a privilege to work with these people. I would like to thank all the members of the group, and our research fellows Carl Craver, Stathis Psillos, Jonathan Schaffer and John T. Roberts. Thanks to Thomas Grundmann and the Emmy Noether Group for making my arrival in Köln a pleasant one.

I am also indebted to Wolfgang Spohn who supported my project from its early days. I have gained a great deal from discussing various topics with him and Michael Baumgartner, Luke Glynn, Franz Huber, Marcel Weber and others in Konstanz.

Perhaps the most intense period of my PhD was spent in Bristol. I would like to thank all the philosophers in Bristol for providing such an intellectually and socially stimulating environment! Especially, I would like to mention Alexander Bird, Chris Clarke, Matt Farr, Stavros Ioannidis, James Ladyman, Hannes Leitgeb, Elina Pechlivanidi, Richard Pettigrew, Giulia Terzian and Emma Tobin.

I was able to improve my arguments through regular discussion with Holly Anderson, Andreas Bartels, Laura Franklin-Hall, Christopher Hitchcock, Phyllis Illary, Meinard Kuhlmann, Marc Lange, Barry Loewer, Tim Maudlin, Stephen Mumford, Julian Reiss, Federica Russo, Samuel Schindler, Gerhard Schurz, Michael Strevens, Jon Williamson and James Woodward. Looking back, I am grateful to the advisors of my Magister-thesis: Holm Tetens (for teaching me to be suspicious of scholastic debates) and Michael Esfeld (for encouraging me to take metaphysics of science seriously). Also thanks to the philosophers in Pittsburgh and the Center for Philosophy of Science, where I have revised the manuscript. Special thanks to Maria Kronfeldner. I realize now how much my parents and my sister have contributed to my work over the years and already during my studies. I was, indeed, so fortunate that they provided a stable environment I could rely on. Sascha Becker, Dennis Kirchhoff and Markus Schrenk: thanks for being there! Two of my chapters rely on previously published work. Chapter 4 uses the main parts of 'Getting Rid of Interventions' (2012), *Studies in the History and Philosophy of Science*. Part C, online first: http://dx.doi. org/10.1016/j.shpsc.2012.05.006, and Chapter 5 reuses 'A Theory of Non-Universal Laws' (2011), *International Studies in the Philosophy of Science*, 25: 97–117. The publishers and I would like to thank Elsevier and Routledge for permission to use the material in the book.

ALEXANDER REUTLINGER

# Part I
# Causation in the Special Sciences and the Interventionist Theory of Causation

# 1

# Causation in the Special Sciences

The aim of this book is to provide a theory of causation in the special sciences (that is, a theory of causation in the social sciences, the biological sciences and other higher-level sciences). I attempt to reach this goal by pursuing a negative target and a positive target: the main negative target is to argue against a currently influential theory of causation, the interventionist theory of causation. The book will focus mainly on the interventionist theory developed in James Woodward's (2003) book *Making Things Happen*. Counter-arguments against the interventionist theory will attempt to show that the central concept of the interventionist theory of causation – that is, the concept of a possible intervention – is immensely problematic. For this reason, I will argue that the interventionist theory of causation is not tenable. The main positive target of the book consists in replacing the interventionist approach with my own explication of causation in the special sciences, the *comparative variability theory of causation*.

In understanding the project, it is crucial to clarify what precisely is meant by 'providing a *theory* of causation'. In current philosophy of science, there are several different philosophical projects – and corresponding questions and tasks – called 'theory of causation':

- first, the semantic project focuses on the truth conditions of causal statements and the analysis of causal concepts (e.g. Lewis 1973b; Woodward 2003);
- second, the metaphysical project is concerned with locating the truth-makers of causal statements in the (physical) world (e.g. Dowe 2000; the contributions in Price and Corry 2007);
- third, the methodological project consists in a rational reconstruction of those methods for scientific testing of causal statements that

are used in the special sciences (for influential works on methodology, cf. Cartwright 1989, 1999, 2007; Reiss 2008; Russo 2009; for causal modelling approaches to causal methodology, cf. Spirtes et al. 2000; Pearl 2000; Williamson 2005).

It is common to distinguish these three projects and it has been emphasized by influential voices in the debate on causation how important it is to be aware of the ambiguity between talking of a 'theory of causation' and asking the corresponding question 'What is causation?' In his classic book *The Cement of the Universe*, Mackie famously draws the distinction between these three philosophical investigations of causation as follows:

> My treatment is based on distinctions between three kinds of analysis, *factual*, *conceptual*, and *epistemic*. It is one thing to ask *what causation is 'in the objects'*, as a feature of the world that is wholly objective and independent of our thoughts, another to ask *what concept (or concepts) of causation we have*, and yet another to ask *what causation is in the objects so far as we know it and how we know what we know about it*. (1980: viii–ix, emphasis added, similarly pp. 1–2)

More recently, Phil Dowe distinguishes a semantic and a metaphysical project:

> As it is the case with many philosophical questions, our question 'What is causation?' is ambiguous, and consequently the philosophy of causation legitimately involves at least two distinct tasks. [...] We begin by considering these two approaches to the task of philosophy. The first is *conceptual* – to illucidate our normal concept of causation. The second is *empirical* – to discover what causation is in the objective world. (2000: 1)

I agree with Mackie and Dowe that the question of what causation is needs to be disambiguated by clarifying whether one pursues the conceptual, the metaphysical, or the methodological project with respect to causation. I assume that these projects neither exclude each other nor are they entirely independent; rather, they should be regarded as being complementary modes of philosophical research. However, in this book attention will be restricted to the semantic project, with the exception, in Chapter 6, where a metaphysical question about causation in the special sciences is addressed. The reason for addressing a metaphysical question is that interventionists claim that their proposed truth conditions for causal statements, and especially their key concept of an intervention, does substantial work for describing the nature of causation.

The method of conceptual analysis has often been associated with clarifying the meaning of everyday discourse and testing the result of these analyses by confronting them primarily (or even exclusively) with our common-sense intuitions. Yet, the conceptual, or semantic, project with respect to causation is not restricted to everyday concepts: especially in philosophy of science, one wants to know how causal notions are used in scientific contexts by scientists. In philosophy of science, analysing the concept of causation and providing truth conditions for causal statements is best reconstructed as an explication in the sense introduced by Rudolf Carnap (Carnap 1950; also Quine 1960: s. 53). In his *Logical Foundations of Probability*, Carnap characterizes an explication as follows:

> The task of an *explication* consists in transforming a given more or less inexact concept into an exact one or, rather, in replacing the first for the second. We shall call the given concept (or the term used for it) the *explicandum*, and the exact concept proposed to take the place of the first (or the term proposed for it) the *explicatum* (1950: 3, original emphasis).

According to Carnap (1950: 7), an explication of, for instance, causation is adequate if and only if (iff) it conforms to the following criteria to a satisfying degree:

1. The explicatum has to apply to paradigm cases of the explicandum;
2. The explicatum has to be stated in an 'exact form'; that is, in a 'well-connected system of scientific concepts' (which include concepts such as laws of nature);
3. The explicatum has to be (potentially) fruitful for empirical scientific research;
4. The explicatum should be as simple as possible; that is, it should be as simple as the more important requirements 1, 2, and 3 permit.

The goal in this book is to provide an explication of causation that is adequate for the special sciences. In particular, I want to test and challenge the adequacy of the currently very influential and widely received interventionist theories of causation (Pearl 2000; Hitchcock

2001; Woodward 2003) as an explication of causation in the special sciences. Focus will be mainly on James Woodward's interventionist theory of causation, because it is the most influential philosophical account of causation in an interventionist vein. To present only a few paradigmatic and well-known examples of the success and fruitfulness of interventionist theories of causation in other areas of philosophy, the interventionist theory of causation has been applied:

• in the debate on mechanistic explanation: Glennan (2002), Craver (2007), Weber (2008);
• in philosophy of biology: Waters (2007);
• in the debate on mental causation (and higher-level causation, in general): Campbell (2007), Shapiro and Sober (2007), and Shapiro (forthcoming);
• in the debate on metaphysics of causation: Eagle (2007), Menzies (2007); and
• in the debate on laws and explanation in the special sciences: Woodward and Hitchcock (2003), and Leuridan (2010).

Moreover, as will be shown in detail, interventionist ideas are popular among social scientists and among philosophers of social science. It is presumably no exaggeration to say that interventionist theories of causation are widely used in philosophy of science. One might even be tempted to call interventionism the new orthodoxy in this area.

So, what do interventionists claim with respect to the truth conditions of causal statements? Interventionists hold, roughly, that $X$ is a cause of $Y$ iff there is a possible intervention on $X$ that changes $Y$. An intervention is a manipulation of the cause – and only of the cause – which is supposed, in principle, to be possible; that is, an intervention is not restricted to the capacities of human agents to intervene. It should be emphasized once again that the goal of this book is, primarily, to discuss interventionist theories and, ultimately, to replace them with a theory (the comparative variability theory) that, it will be argued, is a better candidate for the job. For these purposes, the examples of causal statements are restricted to paradigmatic cases in the social sciences. This restriction is motivated by two reasons:

• interventionists often use examples from the social sciences, especially from economics, to show the adequacy of their theories; and
• social scientists themselves tend to adopt interventionist theories of causation.

In other words, if one takes interventionism to be an explication of causal concepts that are used in paradigmatic causal statements in the social sciences, one makes a maximally strong and fair case for interventionist theories of causation. So, let us begin with a list of paradigmatic causal statements in the social sciences:

*Statement 1*: Caldwell's model of child survival in developing countries formulates the causal claim that a mother's education $C_1$ (measured in number of schooling years) and a father's socio-economic status $C_2$ (measured by income of the fathers) are causes of child mortality $E$ (cf. Russo 2009: 26–30);

*Statement 2*: Low GNP per capita causes a high rate of infant mortality (Little 1998, 2000, cited in Bartelborth 2007: 136);

*Statement 3*: An increase in supply of a commodity – while the demand for it stays the same – causes a decrease in the price (Cartwright 1989: 149f; Kincaid 2004: 177);

*Statement 4*: 'In 1973 OPEC drastically reduced the amount of oil that it supplied to the world, and in short order there were long lines at the gas pumps in the USA and a much higher price for gasoline. When the gasoline price rose, sales of large gas-guzzling automobiles declined. At the same time, exploration of new oil sources increased and eventually the known world oil reserves remained steady, despite earlier widespread predictions that oil reserves would soon be depleted' (Kincaid 2004: 168);

*Statement 5*: The growth of money in an economy has the capacity to raise the general level of prices (Reiss 2008: 266f.);

*Statement 6*: 'Changes in real income cause changes in the money stock' (Hoover 2001: 46);

*Statement 7*: The Iraqi invasion of Kuwait caused the 1990/91 recession in the United States (cf. Hoover 2001: 1);

*Statement 8*: Low social status causes poor health conditions such that the higher one's social status, the better one's health (cf. Cartwright 2009: 411);

*Statement 9*: An increase in the inflation rate causes a decrease in the unemployment rate (cf. Reiss 2008: 169, 192);

*Statement 10*: Commercial hacienda systems tend to agrarian revolt, and plantation systems tend to lead to labour reforms (cf. Kincaid 1996).

Before discussing which criteria of adequacy these paradigmatic causal claims suggest, three disclaimers should be added with respect to these examples of causal statements.

• The book will not be concerned with questions about ontological and explanatory reductionism: in the social sciences and their philosophy, issues of reduction are often discussed under the label 'methodological individualism' (cf. Heath 2005, Kincaid 1996, Hedström and Ylikoski 2010). Neither the related micro–macro distinction will be discussed, nor the exact nature of social macro-entities investigated. Similarly, the questions as to whether these macro-entities can be understood to be causally related at all and whether there is 'downward causation' from the macro-level to the micro-level will not be addressed in this book (cf. Mitchell 2009, Mayntz 2002, Hoover 2001: ch. 5, and Kincaid 1996, 2009: s. 1, for a discussion). These issues must be left aside, as they are outside the scope of this book. However, neither interventionist theories of causation *by themselves*, nor the comparative variability theory of causation *by itself* (see Chapter 8), preclude that macro-events or macro-facts can stand in causal relations (be it in causal relations to other macro-entities or to micro-entities).

• Due to its connection with the micro–macro distinction, no attention will be given to the vast literature on a practical problem for social scientists and decision-makers in politics. Causal relations among social macro variables (such as general price level, inflation rate, real interest rate, GNP, unemployment) sometimes fail to hold once politicians interfere with the social system in question, because the agents (on the micro-level) change their expectations and their decisions due to their awareness of the political interventions (for a classic formulation of this problem, see Lucas 1976, Cartwright 1989: 154f.).

• In the recent debate on causation (cf. Collins et al. 2004a), philosophers have discussed whether absences can be causes. For instance, we often encounter causal statements such as 'you did not keep the promise to water the plant, this caused ...' and 'the World Bank did not regulate the price, this was a cause of ...'. The question some philosophers ask is whether this manner of speaking commits us to accept that, literally, nothing (i.e. the absence of an event) can be a cause. This is mainly a metaphysical dispute about the truth-makers of absence statements. No attempt will be made to solve this metaphysical problem. However, I believe that the interventionist theory of causation, as well as my own approach (see Chapter 8), is compatible with the view that absence statements can be used in causal statements.

Let us now refocus on the list of paradigmatic examples of causal statements in the social sciences. The questions of central importance are: Which semantic interpretation, which truth conditions, do social scientists usually associate with these paradigmatic causal statements? What is the concept of causation that social scientists implicitly or explicitly use in the context of these paradigmatic causal statements? It may not be surprising that social scientists and, especially, economists have not reached perfect agreement in answering these questions. What many economists and, in general, social scientists believe about causation can be divided into two camps: that of the probabilistic theory of causation and that of the interventionist theory of causation.

*Probabilistic theory of causation*: Many working social scientists belong to this camp. According to the paradigm of Granger-causality, $X$ is a cause of $Y$ iff, roughly, $X$ makes a probabilistic difference for $Y$ conditional on the entire past of $X$ (cf. Granger 1962, 1969; a similar approach among philosophers is Suppes 1970). In this case, X makes a probabilistic difference for $Y$ if the conditional probability of $Y$ given $X$ and the past of $X$ does not equal the (unconditional) probability of $Y$ alone. The main goal of this approach is to provide an analysis of causal notions and truth conditions of causal statements in non-causal terms and, indeed, the concepts of conditional probability and probabilistic difference-making are not causal concepts. In this sense, probabilistic theories of causation are conceptually reductive theories. Despite the popularity of probabilistic approaches, the attempt to explicate causation in terms of conditional probabilities has been subject to an increasing number of criticisms among social scientists (Hoover 2001: 13–22, s. 7.2) and philosophers (cf. Hitchcock 2010 for a survey to these problems). Most of the objections complain that probabilistic approaches fail to account for (a) the asymmetry of causation, and (b) the validity of the common-cause principle presupposed by many probabilistic theories of causation. According to this principle, it is the case that if $X$ makes a probabilistic difference to $Y$, then either $X$ causes $Y$, or $Y$ causes $X$, or $X$ and $Y$ are the effects of a common cause (cf. Reichenbach 1956). Conceptually reductive probabilistic theories face additional problems because they fail to avoid Hesslow's paradox and Simpson's paradox, according to which conditional probabilities simpliciter cannot represent any set of causal relations correctly (cf. Skyrms 1980b; Cartwright

1983, 1989; Hoover 2001). To avoid any misunderstanding, none of these criticisms conclusively refutes probabilistic theories of causation.

*Interventionist theory of causation*: The problems of conceptually reductive probabilistic accounts have driven many social scientists (and philosophers of the social sciences) to convert to another camp of theories of causation. Here, causation is explicated by employing the notions of hypothetical intervention, control and manipulation. It is an obvious feature of this second approach to causation that the conceptually reductive spirit of the probabilistic explications is no longer maintained, since the notions of hypothetical intervention, control and manipulation are themselves causal concepts. (Whether or not giving up the conceptually reductive character of explication is problematic will be the subject of detailed discussion in Chapter 7.)

Let us now turn to a more detailed description of the attempt to analyse causation in terms of intervention, control and manipulation. The economist and philosopher Kevin Hoover describes the interventionist idea as follows:

> A causes B if control of A renders B *controllable*. A causal relation, then, is invariant to interventions in A in the sense that *if someone or something can alter the value of A the change in B follows in a predictable fashion.* (1988: 116, emphasis added)

In the same vein, the econometricians Thomas Cook and Donald Campbell write:

> *The paradigmatic assertion in causal relationships is that manipulation of a cause will result in the manipulation of an effect.* This concept of cause has been implicit in all the foregoing examples, and philosophers claim that it reflects the way causation is understood in everyday language. *Causation implies that by varying one factor I can make another vary.* (1979: 36, emphasis added)

The statistician David Freedman distinguishes three uses of regressions in statistics in the social sciences: to summarize data, to predict values of the dependent variable, and to predict the results of interventions. Freedman reserves the notion of cause to the third use:

> Causal inference is different, because a *change in the system* is contemplated; for example, there will be an *intervention*. Descriptive statistics tell you about the correlations that happen to hold in the

data; *causal models claim to tell you what will happen to Y if you change X.*
Patterns in the data are deemed causal because they are useful for
the prediction of the results of policy interventions. (1997: 62,
emphasis added)

In his widely received book *Causality in Macroeconomics*, Kevin Hoover
ties causation conceptually to controllability and interventions:

> A directly causes C if and only if there is some causal path connect-
> ing A to C. The direction of direct causation, the property of causal
> asymmetry, is part of the essential notion of causality as a disposi-
> tional property, and is reflected in the analysis of the causal relation-
> ship in terms of counterfactual conditionals. It is this conditional
> nature that permits causal relationships to be used to control things.
> Control is often the source of our pragmatic interest in causal rela-
> tions, but control *per se* is not part of the concept of causality itself,
> *except insofar as hypothetical control is one way to understand the mean-
> ing of counterfactual conditionals.* (2001: 69, emphasis added)

A similar and influential approach is adopted by David Hendry who
defines causation as invariance under interventions – Hendry uses the
notion of superexogeneity (cf. Engle et al. 1983; and Hendry 2004).

These interventionist accounts proposed by the second camp of social
scientists are broadly in agreement with interventionist theories of causation advocated by philosophers such as Woodward. In other words,
the way that social scientists of the second camp think about causation
themselves seems to cohere with prima facie, and to be in support of,
interventionist theories of causation as presented in philosophy of
science.

In the remainder of this chapter, two criteria of adequacy are
introduced for an explication of causation in the social sciences (the
*naturalist* criterion and the *distinction* criterion). It will be argued that
these standards generalize to the explication of causation in other special sciences besides the social sciences – most importantly, the life sciences. The chapter closes with the outline of the book.

## Explicating causation in the special sciences: Criteria of adequacy

The explication of causation requires criteria of adequacy; that is,
norms according to which we judge whether the explication is apt and

satisfactory. The *naturalist* criterion and the *distinction* criterion should be considered: let us relate them to Carnap's original criteria (p. 5).

*Naturalist criterion*: This is a criterion of adequacy according to which an explication has to apply to paradigmatic causal claims in the sciences, in the sense that the explication captures important features of causation (e.g. causal asymmetry, time-asymmetry) and correctly describes various kinds of causation (e.g. type-level causation, actual causation). This is identified as the naturalist criterion because the features and kinds of causation, which are to be respected, depend on the scientific representations and the scientific practice in a special science. As opposed to this naturalist stance, asking which features speakers assign to causation and which kinds of causation they presuppose in everyday discourse is not the target of this book.

How does the naturalist criterion relate to Carnap's criteria for the adequacy of an explication? In order to answer this question, let us go through each of Carnap's criteria. Carnap's first criterion demands that the explicated concept (the *explicatum*) has to apply to paradigm cases of the concept that is supposed to be explicated (the *explicandum*). The naturalist criterion meets the first of Carnap's criteria because (a) paradigm cases of causal statements, (b) paradigmatic features of causation, and (c) paradigmatic kinds of causation are provided by analysing case studies from successful sciences.

Carnap's second criterion requires that the explicatum has to be stated in an 'exact form'. An explication is given in an exact form if it uses a 'well-connected system of scientific concepts' (such as laws of nature). The naturalist criterion conforms to and, at the same time, restricts Carnap's second criterion because the decision as to which 'scientific concepts' are employed in the explication is also guided by an analysis of the case studies of paradigmatic causal claims in the special sciences.

According to Carnap's third criterion of adequacy, the explicated concept has to be (potentially) fruitful for empirical scientific research. Although it is not the main focus of the semantic project of providing truth conditions for causal statements in the special sciences, the explication of causal concepts which are discussed in this book can claim that the results of these projects are potentially fruitful for scientific research in two ways. First, semantic theories of causation provide, at best, a conceptually clearer picture of a scientific practice (i.e. the search for causes in the special sciences) that is already acknowledged to be empirically successful. Second, some of the accounts of truth conditions for causal statements (e.g. interventionist theories of causation) use the framework of causal modelling. As will be shown in Chapter 2, one goal of causal modelling techniques is normative: causal modelling includes recipes for inferring causes from statistical data – and scientists should adopt these methodological recipes in their research. For this reason, the success of causal modelling would render explications of causation indirectly relevant for empirical research, if these explications were also to use the framework of causal models.

Carnap's fourth criterion requires that the explicatum should be as simple as possible. Let us take simplicity, minimality, or parsimony to be a criterion that is implicitly assumed in any kind of conceptual analysis or explication. In sum, the naturalist criterion captures the heart of Carnap's criteria of adequacy.

The naturalist criterion helps to identify the class of paradigmatic cases of causal statements. This criterion is naturalist because case studies from the special sciences decide which actual usage of causal concepts is a paradigm case. In other words, the purpose of the naturalist standard is to select important features and kinds of causation according to the actual use of causal concepts in the special science. The restriction to the actual application of a concept may be important for analysing the concept of causation. Still, in order to achieve a better understanding of a concept, such as the concept of causation, one would wish to know more than that. One would also wish to inquire the application of causal concepts to sometimes merely possible scenarios. In particular, interest lies in the application of causal concepts as they are actually and paradigmatically used in the special sciences to possible causal scenarios. For this reason, we introduce a second criterion of adequacy for the explication of causation: the distinction criterion.

*Distinction criterion*: According to this criterion, the explicated concept has to be able to distinguish correctly between intuitively different and intuitively possible causal structures. The general motivation for this criterion consists in the idea that the application of concepts should not only be tested for actual cases but also for (merely) possible cases, which would be classified to be cases in which the concept should apply. For instance, one might want to know whether the analysis distinguishes correctly between causal chains (Figure 1.1 illustrates a causal chain from $A$ to $C$ through $B$) and common cause scenarios (Figure 1.2 illustrates that $A$ and $C$ have a common cause, $B$). These causal scenarios will be illustrated using causal graphs. Causal graphs are representations of causal relations, in which capital letters such as $A$, $B$, $C$ ... represent causal relata, and arrows represent causal relations such that the head of the arrow points to the effect (causal graphs and causal models will be discussed in greater detail in Chapter 2).

$$A \longrightarrow B \longrightarrow C$$

*Figure 1.1*    Structure of a causal chain

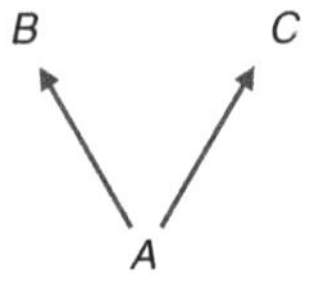

*Figure 1.2*    Structure of a common cause scenario

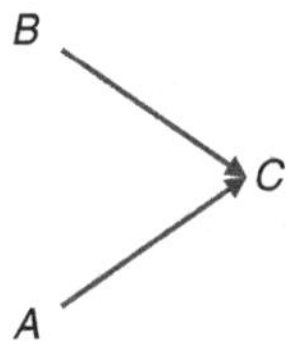

*Figure 1.3*    Structure of a pre-emption scenario

Another well-known class of causal scenarios consists in cases of so-called pre-emption. These are situations in which the effect is overdetermined by several causes; that is, each of the causes is sufficient for the effect. In Figure 1.3, *C* is over-determined by *A* and *B*.

This class of counter-example represents a particular threat to those theories of causation that endorse the claim that causes are necessary for their effects – such as Lewis's counterfactual theory of causation (see pp. 60–7). Scenarios of pre-emption are a threat to theories of causation, assuming that causes are necessary for their effects, because each cause (such as *A* in Figure 1.3) is not necessary for the effect since other causes (such as *B* in Figure 1.3) are sufficient to bring about the effect (various scenarios of pre-emption, which mainly differ concerning the temporal relations among *A*, *B*, and *C*, will be introduced and discussed in Chapter 2).

Of course, not any possible scenario to which the concept in question may be applied has to be assigned equal weight according to the distinction criterion. The danger is that explication becomes, as Tim Maudlin puts it, 'the analysis of fantastical descriptions produced by philosophers, and, not surprisingly, these fantastical descriptions will have in them whatever features the philosophers decided to put in them' (Maudlin 2007: 188). I agree with Maudlin, and others such as

Ladyman and Ross (2007: ch.1), insofar as putting too much weight on the distinction criterion results in esoteric and scholastic debates about 'fantastical' counter-examples to a particular theory of causation, while these alleged counter-examples are warranted only by the (possibly even diverging) intuitions of the philosophers in the debate.

However, taking the worries of naturalist philosophers such as Maudlin and also Ladyman and Ross seriously should remind us of the fact that the distinction criterion can be (and, actually, sometimes is) abused. Naturalist philosophers, with whom I concur, remind us that, for instance, the creation of possible causal scenarios ought to be constrained by being consistent with the best contemporary sciences. Nevertheless, potential abuse and the potential violation of naturalist constraints (e.g. to be consistent with present-day science) should not preclude that we can accept this distinction criterion as a criterion of adequacy for explication.

Let us turn to the naturalist criterion in order to determine what exactly this criterion requires for an explication of causation in the special sciences. The distinction standard will be dealt with in the presentation of Woodward's interventionist theory of causation and its advantages in a more detailed manner (see pp. 57–70). Unfortunately, one cannot simply read off the paradigmatic examples of causal statements in the social sciences to learn which important features of causal statements have to be accounted for by an explication. Yet, by studying the work of social scientists on causation, one can at least acknowledge that social scientists place a high value (a) on distinguishing the following kinds of causation, and (b) on assigning the following features to causal relations (cf. Hicks 1979; Klein 1987; Goldthorpe 2001; Mayntz 2002; Granger 2007: 289, 293; Hedström and Ylikoski 2010).[1] A requirement for any theory of causation in the social sciences is to account for those kinds of causation to which social scientists are committed, and to explain why causation possesses those features that social scientists typically attribute to causal relations. Let us now consider various kinds of causation that social scientists assume.

*Type-level and actual causation*: The special sciences, by and large, are interested in type-level causes as well as actual causes (the latter concerns the historical special sciences). Type-level causation and actual causation differ with respect to the relata standing in causal relations: type-level causation relates event-types; actual causation relates token-events. That is, type-level causal claim are of the form 'type-*C* events cause type-*E* events' while claims about actual causation are of the form 'the event-token *c* causes the event-token *e*'.

A typical example of a type-level causal claim would be: an increase in supply of a commodity causes a decrease in its price. Now, consider a claim about actual causation from the paradigmatic causal claims listed above:

> In 1973 OPEC drastically reduced the amount of oil that it supplied to the world, and in short order there were long lines at the gas pumps in the USA and a much higher price for gasoline. When the gasoline price rose, sales of large gas-guzzling automobiles declined. At the same time, exploration of new oil sources increased and eventually the known world oil reserves remained steady, despite earlier widespread predictions that oil reserves would soon be depleted. (Kincaid 2004: 168)

Another example of a statement about actual causation is: 'the Iraqi invasion of Kuwait caused the 1990/91 recession in the United States' (Hoover 2001: 1).

*Deterministic and probabilistic causation*: According to the examples given for type-level causal claim, some cases of causation are deterministic while other causes are probabilistic (i.e. the causes make a probabilistic difference to the effect).

The examples given often appear to be deterministic, non-probabilistic causal statements; for example: 'changes in real income cause changes in the money stock'. However, other paradigmatic causal statements have an indeterministic character such as: 'commercial hacienda systems *tend to* agrarian revolt, and plantation systems *tend to* lead to labor reforms' (cf. Kincaid 1996, emphasis added). 'Tending to' is here understood as a high positive correlation between commercial hacienda systems and agrarian revolt.

*Multiple causes and degrees of causal influence*: Social scientists believe that real-life cases of causation usually involve multiple causes of a phenomenon, and these multiple causes contribute in different degrees to the effect.

For example, according to Caldwell's model of child survival in developing countries, a mother's education $C_1$ (measured in years of schooling) and a father's socio-economic status $C_2$ (measured by income of the father) are two causes of child mortality $E$ (cf. Russo 2009: 26–30). Additionally, social scientists typically hold the view that multiple causes have *different degrees of causal impact* on the outcome. This fact is often expressed by so-called regression equations or structural equations. In economics and the social sciences, these equations usually take the following linear form: $Y = \alpha X + \beta Z$, with $\alpha$, $\beta$ as parameters representing the strength of influence that $X$ and $Z$ have on $Y$ (Cartwright 1989: ch. 1; Pearl 2000: 27 and ch. 5; Hoover 2001: ch. 2).

Let us now turn to features of causation that social scientists associate with each kind of causation.

*Modal relation*: Causes enforce and produce their effects; that is, some kind of modal connection obtains between cause and effect.

*Time-asymmetry of causation*: Causes precede their effects in time, and not vice versa.

*Causal asymmetry*: The causal relation is asymmetric; that is, if $X$ causes $Y$, then $Y$ does not cause $X$.

*Distinction of spurious correlation and genuine causation*: There is an important distinction between (a) the case of a correlation between $X$ and $Y$, and (b) the case of a causal relation between $X$ and $Y$ (i.e. either $X$ causes $Y$ or vice versa, or the correlation obtains due to a common cause $Z$).

*Context*: Many of these examples of causal statements are (implicitly) believed to hold only on the assumption that other variables are held fixed. This assumption is often referred to by the expression 'ceteris paribus'.

A paradigmatic example of this feature of causation in the social sciences is: 'An increase in supply of a commodity – *while the demand for it stays the same* – causes a decrease in the price' (Cartwright 1989: 149f; Kincaid 2004: 177).

This is not an exhaustive list of features of causation; however, social scientists seem to converge on the criteria in this list, and, consequently, such features of causation should be acceptable, at least, as a guide to the successful explication of causation in the social sciences.

Yet another feature of causal relations has to be added to the list: the no-universal-laws requirement.

*No-universal-laws requirement*: Since there are no strict universal laws in the social sciences, an explication of causation for these disciplines cannot refer to such laws.

The reason for mentioning this criterion separately stems from the fact that it partly differs from the other features on the list: on the one hand, this requirement is on equal footing with the other criteria: social scientists do not claim that the truth conditions of their causal claim depend on universal laws; neither does it make much sense to interpret them in this way in cases where such a commitment is not articulated explicitly. On the other hand, a main reason to add the no-universal-laws requirement to the list of features of causation in the social sciences lies in the history of the philosophical debate about causation and laws in physics and in the special sciences.

To expand on the latter point, in the current debate it is a commonplace among leading philosophers of the social sciences that generalizations in the social sciences are substantially different from statements that are traditionally classified as law statements by philosophers (cf. Cartwright 1983, 1989, 1999; Kincaid 1996, 2004; Hausman 1988, 1992, 1998, 2009; Beed and Beed 2000; Hoover 2001; Roberts 2004; Reiss 2008; Rosenberg 2009; Kincaid and Ross 2009; Hedström and Ylikoski 2010).

If we assume that there are successful and autonomous causal explanations provided in the social sciences, this assumption poses a serious problem for many common philosophical theories of explanation. The most difficult problem in this context consists in the tension between (a) the assumption that explanations involve universal laws of nature, and (b) the plain fact that social scientists – contrary to physicists – cannot rely on such laws, if they explain something.[2] Generalizations in the social sciences – such as the law of demand stating that, if the demand increases and the supply remains constant, then the price increases – are believed to be *lawish* (this is a term of art, see Chapter 5), although they are not universal generalizations. Special science generalizations are lawish in the sense that they play a similar role in scientific practice as laws in fundamental physics: they are important because they are statements used to explain and to predict phenomena, they provide knowledge on how to successfully manipulate the systems they describe, and they support counterfactuals (Mitchell 2000 characterizes generalizations in the special sciences as 'pragmatic laws' by virtue of performing at least one the listed roles). However, traditionally (cf. Lewis 1973a: 73–6; Armstrong 1983: 88–93),[3] the most important feature of a law in order to understand its lawlikeness is universality. Furthermore, picturing lawlikeness mainly in terms of universality has led many theories of causation and explanation to rely on universal laws. It transpires that this assumption is problematic for our purposes: the major challenge for any theory of non-universal laws in the special sciences is to account for their apparent lawish function (in the sense introduced). If we are not able to provide an explication of non-universal laws, then (at least) the philosophy of the special sciences faces a severe problem concerning causation and explanation in its domains. Many theories of causation and explanation in their standard form presuppose universal laws of nature.[4]

The following illustration explains how most theories of causation in their standard form refer, essentially, to laws:

- *Regularity theories* of causation hold that an event *A* causes an event *B* iff:
  - ○ *A* precedes *B* in time;

o  *A* is spatio-temporally connected to *B*; and
o  the proposition that *B* occurs follows logically from the proposition that *A* occurs and a law of nature relates *A* and *B* (cf. Mackie 1980; Baumgartner 2008; Strevens 2009);
- A proponent of the *counterfactual theory*, according to David Lewis (cf. Lewis 1973b), decides whether there is a causal connection between two actual events *A* and *B* by evaluating the appropriate counterfactuals:

o  had *A* been the case, then *B* would have been the case; and
o  had not-*A* been the case, then not-*B* would have been the case.

According to possible worlds semantics for counterfactuals, this evaluation proceeds by comparing the actual world to similar possible worlds. The most important criterion of similarity relies heavily on universal, fundamental laws of nature (at least, in the event that the relevant counterfactuals are supposed to be true at the actual world);
- According to the transfer or conserved quantity theory of causation (cf. Dowe 2000), causation consists in the possession or the transfer of a physical magnitude (typically a conserved quantity). The *transfer* or *conserved quantity theory* of causation builds indirectly on laws of nature: the description of e.g. energy transfer (or possession of energy) is stated in the language of a physical theory, which contains universal natural law statements – most importantly: conservation laws.[5]

Similarly, many theories of explanation essentially refer to law statements: The DN-model and the IS-model claim that explanation consists in deductive or inductive logical inference from laws and initial conditions. Theories of causal explanations inherit their reference to law statements from the concept of causation involved (see Lewis 1986b; Woodward 2003; Woodward and Hitchcock 2003; Strevens 2009). Furthermore, Leuridan (2010) argues that received accounts of mechanisms and mechanistic explanation (such as Machamer et al. 2000; Glennan 2002; Bechtel and Abrahamsen 2005; Craver 2007) also rely on lawish generalizations (see Chapter 9, p. 234).

The problem stemming from many theories of causation and explanation consists in a logical tension between three assumptions:

*Assumption 1*:  The special sciences (a) refer to causes in their domains, and (b) provide explanations in their domains;

*Assumption 2*:  It is evident that the special sciences – possibly, in contrast to physics – cannot rely on universal laws;

*Assumption 3*:   Most philosophical theories of causation and explanation – in their standard form – presuppose universal
laws.

This tension can be formulated as the *nomothetic dilemma of causality and explanation* (cf. Pietroski and Rey 1995: 85; Woodward and
Hitchcock 2003: 2):

*First horn*: If it is a plain fact that the special sciences cannot rely on
universal laws (Assumption 2) and if most philosophical theories of
causation and explanation involve universal laws (Assumption 3), then
there is neither causation nor explanation in the special sciences (negation of Assumption 1).

*Second horn*: If there is causation and explanation in the special
sciences (Assumption 1) and if it is evident that the special sciences
cannot rely on universal laws (Assumption 2), then there is causation
and explanation that does not involve universal laws (negation of
Assumption 3).

If we do not want to give up the immensely plausible opinion that
social scientists refer to causes and provide explanations (Assumption 1)
because of purely philosophical reasons, then we are in need of a theory
of non-universal laws. So, there are good reasons to opt for the second
horn of the nomothetic dilemma. Consequently, the no-universal-laws
requirement is a criterion of adequacy for explicating causation in the
special sciences. Interventionist and probabilistic explications of causation (such as Granger-causality in economics) can be interpreted as
attempts to satisfy this criterion of adequacy.

Let us take stock. Various kinds of causation (such as type-level, actual,
deterministic, probabilistic, and degrees of causal influence) have been
presented, and features of causation (such as the modal character of
causation, its asymmetry and time-asymmetry, a distinction in kind
between causes and mere correlations, and the qualification of causal
statements by clauses such as 'while other factors are held constant'
and 'ceteris paribus'). Social scientists explicitly (and, even more often,
implicitly) accept the presented kinds of causation, and they typically
assign features to causation. Two observations can be made here.

These features not only characterize causation in the social sciences,
they also seem to generalize for the adequate explication of causation in
special sciences other than the social sciences. A good reason to believe
that these features and kinds of causation apply to other special sciences
is indicated by the results of two debates in philosophy of biology. In the
debate on causation in biology and the life sciences in general, by and

large, the same crucial features of causation are highlighted (cf. Okasha 2009). The debate concerning explanatory generalizations in biology also suggests that generalizations that play a role in the scientific practice of biological disciplines lack the features of lawhood as traditionally assigned by philosophers (Beatty 1995; Sober 1997; Mitchell 2000, 2002; Rosenberg 2001; Craver 2007. Steel 2007 is an explicitly interdisciplinary study that focuses on biological and social sciences). Rather than understanding generalizations in biology as universal laws, philosophers of biology prefer to explicate them as context dependent, historically contingent, stable generalizations. The important point is that the results of these debates suggest similar characterizing features of causation in other special sciences than the social sciences.

Second, it is also noteworthy that the features of causation given overlap significantly with the characterization of causal relations in a current debate on the metaphysics of causation: many of these features of causation are thought of as explicating the folk notion of causation (cf. Ladyman and Ross 2007: 268f; Norton 2007: 36–8; Ross and Spurrett 2007: 13f.). The main question driving this debate consists in asking whether the existence of causal relations (as claimed by the special sciences) can be reconciled with a worldview that the present fundamental theories in physics suggest. These features of causation and metaphysical assumptions will be further discussed in Chapter 6.

## Outline of the Book

The book is divided into three parts. Part I sets the stage for a discussion about causation in the special sciences. Chapter 1 was mainly occupied with two tasks: presenting a list of paradigmatic causal claims from the social sciences, which led to the elaboration of two criteria of adequacy for an explication of causation in the special sciences (the naturalist criterion of adequacy, and the distinction criterion of adequacy). Furthermore, two semantic interpretations of causal statements were introduced that social scientists themselves endorse in methodological debates (i.e. the probabilistic theory, and the interventionist theory).

Chapter 2 introduces the currently influential interventionist theory of causation. Among the interventionist theories, focus will be given to Woodward's (2003) widely-received interventionist approach. According to Woodward's theory, $X$ causes $Y$ iff, roughly, there is a possible intervention on $X$ that changes the value of $Y$. An argument will be presented that interventionist theories are prima facie in accordance with the criteria of adequacy presented in Chapter 1.

Part II of the book is devoted to objections against Woodward's interventionist theory. As stated in the first section of this chapter, the main negative target of this book is to argue that the central concept of the interventionist theory of causation (i.e. Woodward's approach) – that is, the concept of a possible intervention – is problematic. For this reason, it is argued that the interventionist theory of causation is not tenable. The main positive target of the book consists in replacing the interventionist approach with my own explication of causation in the special sciences, the comparative variability theory of causation (discussed in detail in Part III).

In Chapter 3, it will be argued that interventionists fail to specify truth conditions for interventionist counterfactuals. This is a puzzling fact because interventionist counterfactuals – that is, counterfactuals of the form 'if there were an intervention on $X$, then $Y$ would change' – are the fundamental building block of the interventionist theory of causation. Because of this desideratum, two alternative semantics for interventionist counterfactuals are proposed that are taken to be consistent with the interventionists' intuitions about causation.

Chapter 4 will present two arguments to the conclusion that the interventionists' key notion of a possible intervention is problematic. The trouble with interventionism stems from the modal character of possible interventions: Woodward merely requires that interventions be logically possible. The argument is offered that, if one assumes that interventions have to be merely logically possible, then one runs into the following difficulties: it is either the case that interventions become dispensable for the task of explicating causal concepts and stating truth conditions of causal claims, or interventionist counterfactuals have to be counter-intuitively evaluated as false (even if one uses the semantics for counterfactuals explored in Chapter 3). Woodwardian interventions are not considered.

Chapter 5 will present a problem that ensues from Chapters 3 and 4. The problem arises if one accepts the naturalist criteria of adequacy (see Chapter 1), then any theory of causation for the special sciences has to account for the no-universal-laws requirement. Interventionists respect and, prima facie, satisfy this requirement as a criterion of adequacy: according to the interventionist view, causation requires invariant relationships between cause and effect. These invariant relations are described by so-called invariant generalizations, which are not required to be universal. However, the invariance of a generalization is defined in terms of possible interventions; that is, a generalization is invariant under possible interventions. This is unfortunate for interventionists, because the notion of a possible intervention is a troubled one, as argued in Chapter 4. Chapter 5 pursues the claim that one can account for the no-universal-laws requirement without appealing to the notion of an intervention. A defence is offered for an alternative explication of non-universal laws in special sciences. Chapter 6 turns to a metaphysical claim concerning the features of causation. Interventionists argue that the notion of an intervention enables us to account for various features of causal relations (such as the asymmetry and time-asymmetry of causation). The main goal in this chapter is to reject the interventionists' positive argument for the success of employing the concept of a possible intervention: the open-systems argument. The open-systems argument, which is supposed to explain the time-asymmetry (and other features) of causation, will be discussed. It will be argued that not only is the open-systems argument unsound, but also it crucially depends on the flawed notion of a possible intervention. There is, however, an alternative way to account for the time-asymmetry of causation: the statistical-mechanical account of causation. Chapter 6 is the only section of the book that explicitly addresses the metaphysics of causation, thereby deviating from the main semantic project of the explication of causal concepts.

Part III will pursue the positive aim of the book. It is here that I will advocate an adequate explication of causation in the special sciences: the comparative variability theory of causation.

Chapter 7 addresses the issue that interventionist definitions of causal notions are conceptually circular. However, despite the criticisms presented in Part II, the conceptually non-reductive feature of interventionist definitions will be defended. Being conceptually non-reductive is a feature that is not genuine to interventionist theories of causation. As it will transpire in Chapter 8, my own comparative variability theory of causation shares this feature with the interventionist theory (without, however, relying on the notion of a possible intervention). The purpose of Chapter 7 is to preclude objections against the conceptually non-reductive character of comparative variability theory of causation in the special sciences (only as a side-effect, this chapter also defends interventionist theories in this respect).

Chapter 8 advocates the comparative variability theory of causation, which aims at preserving two virtuous features of interventionist theories that are responsible for their success:

• the truth conditions of causal statements can be stated in terms of non-backtracking counterfactual dependence between effect and cause; and

- the counterfactual dependence of the effect upon the cause obtains only conditional on holding other causes fixed.

The comparative variability approach does not rely on the notion of a possible intervention. It will be argued that this approach is an adequate explication of causation in the special sciences, and, further, that it satisfies the criteria of adequacy in a way that is superior to Woodward's interventionist theory. For this reason, it will be argued that comparative variability theory should replace interventionist theory.

Chapter 9 will explore the consequences of a comparative variability theory of causation with respect to theories of causal explanation, mechanistic models, and the conditional analysis of dispositions.

# 2
# The Interventionist Theory of Causation

## The interventionist explication of causation

In his highly influential book *Making Things Happen*, James Woodward argues for an interventionist theory of causation. The key idea underlying interventionist accounts is that, roughly, *X is a cause of Y iff there is a possible intervention on X that changes Y* (see also Chapter 1). An intervention is a manipulation of the cause – and only of the cause – and it is assumed that the manipulation is, in principle, possible. It is important to observe that an intervention is neither restricted to the capabilities of human agents nor is it the case that the concept of human agency and decision-making is necessary for explicating the concept of causation (in this respect, interventionist approaches clearly differ from agency theories of causation such as von Wright 1974, Menzies and Price 1993, cf. Woodward 2003: 103, 123–7).

Let us consider an application of the interventionist theory of causation. For instance, suppose that the type-level causal claim *drinking coffee causes nervousness* is true. Does this claim tell us more than the fact that drinking coffee is correlated with being nervous? Interventionists think this is, indeed, the case. Interventionists hold that drinking coffee causes nervousness iff there is a possible way to manipulate the coffee consumption (e.g. by serving only fruit juice in cafes) such that, as a result of this intervention, former coffee consumers would be less nervous. It is crucial to emphasize that interventionist theories of causation, especially that of Woodward, are not concerned with the epistemology of causation. The aim of interventionists is to provide a semantic account of causal statements.[1] Woodward is very explicit about the semantic goal of his approach: 'my aim is to give an account of the content or meaning of various locutions, such as *X causes Y* [...]' (Woodward 2003: 38, also pp. 7–9).

After this preliminary sketch, let us examine a more precise account of Woodward's theory. In order to understand Woodward's theory of causation, we should begin with his definition of a direct type-level cause, which will be followed by the most important conceptual tool of the interventionist theory: the notion of a possible intervention. Woodward defines a type-level cause as follows:

**Definition: Direct type-level cause**
A necessary and sufficient condition for $X$ to be a direct cause of $Y$ with respect to some variable set **V** is that there be *a possible intervention on X* that will change $Y$ (or the probability distribution of $Y$) when all other variables are held fixed at some value by interventions (Woodward 2003: 55, also p. 59 as part of definition **M**).

So, according to Woodward, the existence of a possible intervention (given that 'other variables are held fixed') is a necessary and sufficient condition for the application of the concept of a *direct type-level cause*. In fact, according to Woodward's interventionist theory of causation, the existence of a possible intervention is a necessary and sufficient condition for the application of *any* causal notion: if $X$ is a direct type-level cause of $Y$, then there is a *possible intervention* on $X$ that changes $Y$ (note that this also holds for other causal notions besides direct type-level causation, such as actual causation). This definition encodes a special sort of counterfactual dependence between effect and cause, which is expressed in terms of the truth of a so-called interventionist counterfactual. An *interventionist counterfactual* is a conditional of the following form: if there were an intervention such that the value of $X$ were changed to some value $x$, then the value of $Y$ would also change. In this sense, interventionist theories of causation are a special kind of counterfactual theory of causation.

Let us illustrate this definition of a direct type-level cause with an example. Suppose that the paradigmatic causal claim 'an increase in the supplied quantity of a good causes decrease in the price of this good' expresses a direct causal link between supply and price. According to the interventionist definition of direct causation, 'an increase in the supplied quantity of a good directly causes a decrease in the price of this good' is true iff all other variables besides supply and price of the good (e.g. the demand for the good) are held fixed, and there is a possible intervention on the supplied quantity that alters the price. We may express the right-hand side of this bi-conditional in terms of the interventionist counterfactual 'if there were an intervention on the supply such that the supply changes and all other variables besides supply and price (e.g. demand) are held fixed, then the price changes'. Woodward's definition is stated in terms of random variables, such as $X$ and $Y$. Variables have been imported in the larger philosophical debate from statistics, the methodology of causal inference, and theories of type-level causation (i.e. theories about the truth conditions for type-level causal statements). A variable $X$ (in the terminology of statistics) is a function $X: D \rightarrow ran(X)$, with a domain D of outcomes or individuals, and the range $ran(X)$ of possible values of $X$. For quantitative variables $X$, $ran(X)$ is the set of real numbers; qualitative variables are binary; that is, $ran(X) = \{0; 1\}$. On notation: capital letters, such as $X, Y, Z, \ldots$ denote variables; lower case letters, such as $x, y, z, \ldots$ denote values of variables; the proposition that X has a certain value x is expressed by a statement of the form $X = x$; that is $X = x$ is a statement about an event-type (cf. Sinai 1992, Eagle 2011: section 1.9). In other words, we can use the language of variables to express quantitative and qualitative type-level causal claims. For instance, the event-type that an object has a mass of 5.68kg is represented by the statement $M = 5.68$, with $M$ as a quantitative variable representing that an object has mass, and 5.68 as a possible value of $M$. An example of a qualitative event-type represented by a binary variable is: the qualitative event-type *drinking coffee* is represented by the event-type statement $C = 1$, where $C$ is a binary variable, which represents coffee consumption, ranging over the set $\{0;1\}$ of possible values. Obviously, being able to accommodate quantitative language as well as qualitative language (in terms of binary variables) is an advantage for an explication of causation in the (special) sciences.

The language of variables allows us to draw a further distinction between (quantitative and qualitative) type-level statements and statements about actual causation. According to theories of type-level causation, the relata of causal relations are event-types represented by statements expressing the fact that a possible value is assigned to a variable. This deviates from many received philosophical views: many philosophers, such as David Lewis, focus on singular causation and take the causal relata to be events described by event statements such as 'event *e* occurs'. Interventionism copes with actual causation in a form that is derivative of type-level causation: interventionists express singular event statements by, first, a random variable taking a certain value (which constitutes an event-type such as $X = x$), and, second, by claiming that this event-type, expressed by $X = x$, is actually instantiated by an object.[2] For instance, the token-event that the actual temperature in my office is 30°C today at noon is represented by the statement 'the event-type represented by $T = 30°$ is actually instantiated in my office today at noon'.

An essential part of the truth conditions of a causal statement (and of interventionist counterfactuals) is the existence of a possible intervention. What is a possible intervention? Woodward sketches the notion of an intervention as follows:

> *It is heuristically useful to think of an intervention as an idealized experimental manipulation* carried out on some variable $X$ for the purpose of ascertaining whether changes in $X$ are causally related to changes in some other variable $Y$. [...] The idea we want to capture is roughly this: an intervention on some variable $X$ with respect to some second variable $Y$ is a causal process that changes the value of $X$ in an appropriately exogenous way, so that if a change in the value of $Y$ occurs, it occurs only in virtue of the change in the value of $X$ and not through some other causal route. (Woodward 2003: 94, emphasis added)

In other words, a possible intervention on a variable $X$ is something that changes the value of the effect $X$. Interventionists such as Woodward understand this change as being causal. The basic idea of an intervention consists in assuming an additional variable $I$ that causes a local change of the value of $X$; that is, the additional cause $I$ sets a variable $X$ to a certain value. This additional cause $I$ is called an 'intervention' or, more precisely, an 'intervention on a variable'. An intervention variable $I$ on $X$ can be represented by an additional exogenous variable $I$ having the possible values $\{i_1, ..., i_n\}$. Suppose that $X$ directly causes $Y$, because some value $y_1$ of $Y$ counterfactually depends on some value $x_1$ of $X$ and the fact that $X$ has the value $x_1$ is the result of an intervention on $X$ (as Woodward's definition requires). According to Woodward, the fact that $X$ has the value $x_1$ is itself the effect of another cause: it is the effect of the intervention variable $I$ (Figure 2.1 illustrates the idea that $I$ is modelled as an additional variable in a causal graph; the arrows represent causal relevance. Causal graphs will be discussed in greater detail in the next section).

Figure 2.1 was a rough sketch of the idea of an intervention. Let us now turn to a rigorous definition of an intervention. Woodward defines

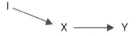

*Figure 2.1*   *I* modelled as an additional variable in a causal graph

interventions more precisely by means of intervention variables. So, let us revisit his definition of an intervention variable:

**Definition: Intervention variable**
*I* is an intervention variable for *X* with respect to *Y* if and only if *I* meets the following conditions:

(I.1) *I* causes *X*.

(I.2) *I* acts as a switch for all the other variables that cause *X*. That is, certain values of *I* are such that when *I* attains those values, *X* ceases to depend on the values of other variables that cause *X* and instead depends only on the value taken by *X*.

(I.3) Any directed path from *I* to *Y* goes through *X*. That is, *I* does not directly cause *Y* and it is not a cause of any causes of *Y* that are distinct from *X* except, of course, for those causes of *Y*, if any, that are built into the *I–X–Y* connection itself: that is except for (a) any causes of *Y* that are effects of *X* (i.e. those variables that are causally between *X* and *Y*), and (b) any causes of *Y* that are between *I* and *X* and have no effect on *Y* independently of *X*.

(I.4) *I* is (statistically) independent of any variable *Z* that causes *Y* and that is on a directed path that does not go through *X*. [...]

(I.5) *I* does not alter the relationship between *Y* and any of its causes *Z* that are not on any directed path (should such a path exist) from *X* to *Y* (Woodward 2003: 98f.).

Let us explore this rather lengthy definition of an intervention variable with the example of a randomized experiment – Woodward (2003: 94–8) uses a similar example for the purpose of illustration. Imagine the following scenario in a medical laboratory: the scientists in the laboratory want to find out whether the new drug Silencin cures tinnitus. In order to do so, they randomly divide the total group of tinnitus patients, all volunteers for the experiment, into two groups: one group receives Silencin (the treatment group), while the other group receives a placebo (the control group). Suppose we represent the experimental set-up with the following binary variables: the variable *I* represents whether a patient is given Silencin or the placebo (*I* = *1* represents a patient given Silencin, *I* = *0* represents a patient given the placebo); the variable *T* represents whether the patient takes the drug prescribed (*T* = *1* represents that the drug is taken, *T* = *0* represents that it is not); the variable *R* represents whether the patient recovers (*R* = *1* represents recovery and *R* = *0* represents no recovery). Intuitively, the variable *I* represents how the scientists in the laboratory intervene and manipulate the patients in the treatment group and the control group. What the scientists would like to know is whether taking Silencin ($T = 1$) makes a causal difference to recovering from tinnitus ($R = 1$). The decisive question for the researchers in the laboratory is how they should act, if they want the test of Silencin to be reliable. Note, again, that Woodward (2003: 95) explicitly emphasizes that the illustration of a randomized controlled experiment is merely heuristic: it is supposed to show intuitively how interventions contribute to understanding the truth conditions of causal statements. He underlines that the methodological question of whether and how one can use randomized experiments to test causal claims has to be distinguished from his own search for truth conditions of causal claims.

So, let us use the question 'How should the scientists act, if they want the experiment to be a reliable test of taking Silencin as a cure for tinnitus?' in order to reconstruct and illustrate Woodward's definition of an intervention variable.

I.1    According to the first condition, the variable $I$ is a cause of $X$. For our scientists, this has a straightforward implication: if they want the test for Silencin to be reliable, then they have to require that their distribution of Silencin and of the placebo is a cause of the patients' ingestion of either of the drugs. For example, the scientists have to exclude that patients spit out the drug or throw it away.

I.2    According to the second condition, there is at least some value of $I$ such that if $I$ takes this value, then $X$ depends only on $I$ and on no other variables; that is, $I$ is the only cause of $X$. The methodological rule for our researchers is: rule out the presence of any other cause of the recovery other than taking the drug (or the placebo). For instance, the researchers have to rule out that the patients take other drugs against tinnitus, and they have to rule out that the personal interaction between researcher and patient has an effect on recovery (e.g. the researcher might be a very kind person and this, in turn, may cause the patient to relax).

I.3    Woodward's third condition states that $I$ is not a direct cause of $Y$, and if $I$ is a cause of $Y$, then $I$ is an indirect cause of $Y$ via a causal path leading through $X$ and a – possibly empty – set of intermediate variables $Z_1, ..., Z_n$. This matters to our researchers in that they have to guarantee that giving Silencin to a patient causes recovery only via the ingestion of the drug – if Silencin causes recovery at all. In other words, the researchers have to exclude, for example, cases where every patient who received Silencin is also administered an alternative and highly successful drug, say Quietin, which cures tinnitus within seconds after its intake.

I.4   Woodward's fourth condition requires that $I$ be probabilistically independent of other causes $W_1$, ..., $W_n$ of $Y$, which are not on a causal path leading from $X$ to $Y$ (cf. Woodward 2003: 99–102). That is, if the researchers want the experiment to be reliable, then they have to exclude, for instance, that the following typical problem occurs: that the patients themselves and the experimenters know whether a patient is in the treatment group or in the control group ('double-blind' randomized experiments were devised to prevent such occurrences). The trouble is that, if a patient knows that he or she is in the treatment group, this knowledge might increase his or her hope of being cured; this knowledge might have a placebo effect. Likewise, whether or not an experimenter knows that a patient is in the treatment group might effect the researcher's observations of whether the patient recovers or not (because the researcher wants the test to be successful, because the test is important for the next application for funding). In order to exclude these correlated causes of recovery, one usually double-blinds randomized experiments by ensuring that neither the patients nor the experimenters (who hand out Silencin to the patients) are aware of who belongs to the control group and who belongs to the treatment group.

I.5   The fifth condition of Woodward's definition of an intervention variable states that $I$ does not alter the relationship between $Y$ and its causes $W_1$, ..., $W_n$, which are not on a causal path leading from $X$ to $Y$. Our researchers in the laboratory are aware of the fact that they are not (always) able to isolate the patients from all other causes of recovery from tinnitus. For instance, suppose that the patients calm down in the laboratory environment because they forget their stressful jobs during the hours in the laboratory. Such stress relief might be another cause for curing tinnitus problems: an increase in stress relief causes a decrease in tinnitus problems. Thus, the situation is that the researchers cannot isolate the effect of Silencin in the sense that they cannot extinguish stress relief as another cause of recovery. However, they can try to minimize the influence (by not designing the laboratory in a too comfortable and cosy style), or they can, at least, attempt to hold these other cause of recovery fixed such that all the patients are in the same (more or less stress relieving) environment.

Having reconstructed and illustrated Woodward's definitions of an intervention variable, let us now proceed to Woodward's definition of an intervention. The notion of an intervention variable is used to define the notion of an intervention:

**Definition: Intervention**
Any value $i_i$ of an intervention variable $I$ (for $X$ relative to $Y$) is an *intervention on $X$ relative to $Y$* iff it is the case that there is a value of $X$ that counterfactually depends on the fact that $I$ has the value $i_i$ (cf. Woodward 2003: 98).

In order to illustrate this definition, recall the example of the randomized controlled experiment for the efficacy of Silencin. In that example, *the variable $I$ is an intervention variable for $T$ relative to $R$*. If we suppose that taking Silencin does cause recovery, then $I = 1$ is an intervention on $T$, because there is a value of $T$ that counterfactually depends on the values of $I$ (including $I$'s value 1). That is, the counterfactuals 'if it were the case that $I = 1$, then it would be the case that $T = 1$' and 'if it were the case that $I = 0$, then it would be the case that $T = 0$' are both considered to be true.

Woodward draws the following heuristic analogy from randomized experiments to the basic idea of the interventionist theory of causation. Suppose that taking Silencin causes a decrease in tinnitus problems. According to the basic idea of interventionism, 'taking Silencin causes a decrease in tinnitus problems' is true iff there is a possible intervention on taking Silencin that alters tinnitus problems. Woodward's heuristic says that to claim there is a possible intervention can be illustrated and made plausible by the claim that there is, in principle, a reliable randomized controlled experiment that confirms Silencin causes a decrease in tinnitus problems. Apply this analogy to one of the paradigmatic causal claims from the social science (see Chapter 1): from an interventionist point of view, 'low GNP per capita causes a high rate of infant mortality' is true iff there is a possible intervention on the GNP per capita (e.g. one that lets the GNP per capita increase) that alters the rate of infant mortality. According to Woodward, the claim that there is a possible intervention on the GNP can be illustrated and made plausible by the claim that there is, in principle – though maybe not feasible for politicians – a reliable randomized controlled experiment that confirms that low GNP causes a high rate of infant mortality.

Moreover, Woodward repeatedly underlines three characterizing features of an intervention.

*Relativity*: An intervention on $X$ is always to be understood as an *intervention relative to Y*, the purported effect (cf. Woodward 2003: 103). An intervention on $X$ relative to $Y$ might *not* be an intervention on $X$ *relative to some variable U* (for $U \neq Y$), because – although condition (a) for being an intervention variable might be satisfied for some variable $I$ – the conditions (b)–(e) might not be satisfied for the second case (for instance, because other variables than in the case of $U$ have to be held fixed as effect variables in the case of $Y$).

*Modal character*: Interventions are not required to be possible in the sense that they are feasible actions for human agents. Rather, the possibility of intervening is pictured here in a modally less restrictive sense. The key idea seems to be that interventionists want to say that, according to their theory of causation, it is, in principle, possible to intervene. Following this line of thought, Woodward (2003: 128) rejects the idea that interventions need to be physically possible; he requires merely that interventions be logically possible. This modal character of interventions will be discussed in Chapter 4.

*Invariance under interventions*: Woodward requires that interventions on $X$ do not alter the functional relationship $Y = f(X)$ between variables. These functional relationships are linguistically represented by so-called invariant generalizations. According to Woodward and Hitchcock (2003: 17) and Woodward (2003: 250), a generalization G is minimally invariant iff the testing intervention condition holds for G. The testing intervention condition states for a generalization G of the form $Y = f(X)$: (a) there are at least two different possible values of an endogenous variable $X$, $x_1$ and $x_2$, for which $Y$ realizes a value in the way that G describes, and (b) the fact that $X$ takes $x_1$ or, alternatively, $x_2$ is the result of an intervention. The invariance theory of laws will be the subject of Chapter 5.

The notion of a possible intervention, interventionist counterfactuals, and the definition of a direct type-level cause are the building blocks of the interventionist theory of causation. Based on these concepts, other causal notions can be defined and will be discussed shortly.

But let us pause for a moment: What is the goal of Woodward's interventionist theory? And, what kind of definitions of causal notions are we facing? As pointed out at the very beginning of this section, Woodward's central aim in *Making Things Happen* 'is to give an account of the content or meaning of various locutions, such as $X$ causes $Y$, $X$ is a direct cause of $Y$, and so on, in terms of the response of $Y$ to a hypothetical idealized experimental manipulation or intervention on $X$' (Woodward 2003: 38). I agree with Strevens (2008: 184) that the best interpretation of Woodward's approach and of his self-declared goal consists in reconstructing Woodward as providing truth conditions for causal claims. Providing truth conditions for causal claims, such as 'X directly causes Y', is supposed to be not merely a description of the actual usage of causal notions in everyday and scientific concepts. Woodward's semantic project is selective and partly revisionary, and it is best conceived as an explication of causal concepts (and as an account of truth conditions of a variety of causal claims) that is constrained by various criteria of adequacy (cf. Woodward 2003: 7–9). Woodward characterizes his definitions of causal notions as follows:

> It is true that M[aking] T[hings] H[appen] presents 'definitions' of the various causal concepts mentioned above. Some of these concepts are defined in terms of others and all are taken to be related in various ways to the notion of an intervention, which I also define in M[aking] T[hings] H[appen]. [...] I also explain how these definitions are to be taken in the opening pages of M[aking] T[hings] H[appen] (pp. 7–9): they are definitions in the sense that, say, a mathematician might define the notion of continuity of a function in terms of the notion of an open neighborhood. Such definitions are to be judged by their usefulness for various purposes – in capturing previous usage, in clarifying notions that were previously unclear and distinguishing them from related but different notions, in establishing fruitful connections with other concepts and so on, rather than in terms of whether they adequately capture fundamental metaphysical relationships. (Woodward 2008: 195f.)

The guiding thought behind this passage, combined with Woodward's (2008: 193–6) insistence that he is not particularly interested in the metaphysics of causation, seems to be that his goal is the explication of causation for scientific (and also everyday) contexts. However, if Woodward's interventionist theory of causation is mainly concerned with truth-conditional semantics for causal claims, then many philosophers are likely to feel bewildered by his proposal: the main reason for this bewilderment is that the interventionist explications of causation are apparently circular (cf. Psillos 2004; Hiddleston 2005; Menzies 2007; Strevens 2007, 2008). The explications are circular because, for instance, the notion of a direct type-level cause $X$ of $Y$ is explicated by means of a possible intervention on $X$ – which in turn is defined as a specifically constrained cause of $X$ – and under the assumption that other causes of $Y$ are held fixed (among these causes are direct type-level causes of $X$).

Woodward (2003: 20–2, 104–7, 2008: 203f.) defends his view by arguing that his definitions are not viciously circular. According to Woodward, a definition of '*X* directly causes *Y*' is viciously circular only if it presupposes that *X* directly causes *Y*. This is not the case for interventionist definitions, he argues. Whether or not this circularity turns out to be a serious problem for the interventionist approach, let me diagnose for now that the kind of explications that Woodward presents are not what many philosophers expect from a semantic theory of causation: many philosophers expect that if one sets out to provide the truth conditions for causal claims, then the truth conditions must not refer to causal facts. In other words, most philosophers expect the analysis, the definition, or the explication of a concept to be *conceptually reductive*. This is a bugging issue for interventionists, and circularity is further discussed in Chapter 7.

A further observation can be made about the peculiar feature of model-relativity that inheres in the interventionist approach. Causal concepts are defined relative to a 'set of variables' or a causal model. Also the truth conditions for causal claims are stated relative to a 'set of variables' or a causal model. It is, however, not evident how these models relate to the world (cf. Strevens 2007, 2008 for a discussion of model-relativity). In other words, it is not evident why one ought to believe that a given causal model is a representation of the world – be it for the reason that the model is true, or that it is isomorphic or homomorphic to the target system that the model is intended to represent. For the purposes of this book, I will bracket the interesting question about which feature renders some models representational. This is a complicated and demanding epistemological problem that (a) is not a unique problem of the interventionist theory (other approaches to causation), and that (b) has been the subject of an extensive debate about scientific realism and the epistemological status of models in the sciences. However, the models to which interventionists refer when they define causal concepts are not immediately in danger of being free-floating imaginations unsupported by evidence, because they are closely tied to methodologies of evidence for causal models. Several approaches to evidential support of causal models will be presented (pp. 45–6). Although it is important to bear in mind that semantic and methodological projects are distinct philosophical enterprises, a complete theory of causation ought to unite semantic, metaphysical, and methodological accounts (see also Chapter 9, p. 250).[3] Let us now return to the interventionist explications of other causal notions besides the notion of direct causation.

Having introduced the notion of a possible intervention and the notion of direct type-level causation, Woodward is able to define further causal notions besides the notion of a direct type-level cause. Other causal notions can be captured in the interventionist framework such as the concepts of an indirect cause, a total cause, and an actual cause. Based on the notion of a direct cause, Woodward defines indirect (contributing) causation:

**Definition: Indirect type-level cause**
A necessary and sufficient condition for $X$ to be a (type-level) contributing cause of $Y$ with respect to some variable set **V** is that (i) there be a directed path from $X$ to $Y$ such that each link in this path is a direct causal relationship; that is, a set of variables $Z_1$, ..., $Z_n$ such that $X$ is a direct cause of $Z_1$, which is in turn a direct cause of $Z_2$, which is a direct cause of ... $Z_n$, which is a direct cause of $Y$, and that (ii) there be some intervention on $X$ that will change $Y$ when all other variables in **V** that are not on this path are fixed at some value. (Woodward 2003: 59)

Let us observe this definition by the simplest scenario of indirect causation. According to the causal graph presented in Figure 2.2, $X$ is an indirect cause of $Y$ via a causal path through $Z$, and $W$ is a cause of $Z$ which is not intermediate between $X$ and $Y$: it is on a different causal path.

According to the first condition of the definition of an indirect type-level cause, each link in the path from $X$ to $Y$ – that is, one link from $X$ to $Z$ and another link from $Z$ to $Y$ – is a link of direct causation. According to the second condition of the definition, the interventionist counterfactual, 'if there were an intervention on $X$ and those causes that are not on the causal path from $X$ via $Z$ to $Y$ (such as $W$) are held fixed, then the value of $Y$ would be different' is true.

Woodward (2003: 50f.) takes the definitions of direct and indirect causation to be the definitions of *contributing* causation. In the case of contributing causation, $X$ has a causal impact on $Y$ *given that* variables

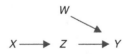

*Figure 2.2*   A simple scenario of indirect causation

that are not on a directed path from $X$ to $Y$ are held constant by interventions. As opposed to contributing causes, *total* causes are *not* relativized to holding fixed other off-path variables; that is, a total cause $X$ has an impact on the effect variable Y regardless of the values that other off-path variables $W_1, ..., W_n$ take. Woodward defines a total, type-level cause as follows:

**Definition: Total cause**
$X$ is a total cause of $Y$ if and only if there is a possible intervention on $X$ that will change $Y$ or the probability distribution of $Y$ (Woodward 2003: 51).

Besides type-level causes, Woodward also defines actual causes. Woodward defines actual causation derivatively of type-level causation. Recall that a paradigmatic claim about actual causation is 'In 1973 OPEC drastically reduced the amount of oil that is supplied to the world, and in short there were long lines at the gas pumps in the USA and a much higher price for gasoline' (Kincaid 2004: 186). According to Woodward, an actual event $c$, described by the event statement 'an object actually instantiates the event-type $X = x$' (or, briefly, '$X = x$ is actually true'), is an actual cause of an actual event $e$, described by the event statement 'an object actually instantiates the event-type $Y = y$' (or, briefly, '$Y = y$ is actually true'), iff the following two conditions (AC*1) and (AC*2) are satisfied:

**Definition: Actual cause**
(AC*1) $X = x$ is actually true and $Y = y$ is actually true. (AC*2) For each directed path P from $X$ to $Y$, fix by interventions all direct causes $Z_i$ of $Y$ that do not lie along P at some combination of values within their redundancy range. Then determine whether, for each path from $X$ to $Y$ and for each possible combination of values for the direct causes $Z_i$ of $Y$ that are not on this route and that are in the redundancy range of $Z_i$, whether there is an intervention on $X$ that will change the value of $Y$. (AC*2) is satisfied if the answer to this question is 'yes' for at least one route and possible combination of values within the redundancy range of the $Z_i$ (Woodward 2003: 84).

In order to understand this definition, we need to elucidate the notion of a redundancy range on which the definition is based (cf. Woodward 2003: 83; Hitchcock 2001: 290). A redundancy range concerns a range of actual and counterfactual[4] possible values that a variable might have, which lies off the directed path P (in question) from $X$ to $Y$ representing the putative cause and effect. In other words, the redundancy range is defined with respect to other direct causes $W_1$, ..., $W_n$ of $Y$ – both (a) causes on a different path P* from $X$ to $Y$ and (b) causes of $Y$ that do not lie on any path that leads through $X$. The values of $W_1$, ..., $W_n$ are in the redundancy range with respect to the path P from $X$ to $Y$ iff the values of $W_1$, ..., $W_n$ do not change the actual values of the variables on path P.

In other words, Woodward's definition of actual causation states that $X = x$ is an actual cause of $Y = y$ iff the conjunction of the following conditions is satisfied:

- *Actuality condition*: $X = x$ refers to the actually occurring event $c$ and $Y = y$ refers to the actually occurring event $e$.
- *Redundancy range condition*: The values of all variables that are not on the directed path P from $X$ to $Y$ are in the redundancy range with respect to the actual value of the variables on path P.
- *Interventionist counterfactual dependence condition*: The interventionist counterfactual 'if there were an intervention such that the value of $X$ were changed from its actual value $x$ to some counterfactual value $x^*$, then the actual value $y$ of $Y$ would change to a counterfactual value $y^{*}$' is true.[5]

Let us observe this interventionist definition of actual causation. The claim about actual causation 'In 1973 OPEC drastically reduced the amount of oil that is supplied to the world, which caused a much higher price for gasoline' is true iff:

- *Actuality condition*: The event statements 'In 1973 OPEC drastically reduced the amount of oil that is supplied to the world' and 'the price for gasoline was much higher' are actually true.
- *Redundancy range condition*: The values of all variables that are not on the directed path P from a variable representing the supply to a variable representing the price are in the redundancy range with respect to the actual value of the price.
- *Interventionist counterfactual dependence condition*: The interventionist counterfactual 'if there were an intervention such that the supply of oil were changed from its actual value to some counterfactual value (for instance, OPEC could have increased the supply drastically), then the price for gasoline would change to a counterfactual value' is true.

Let us briefly take stock. I have reconstructed Woodward's interventionist theory of causation. Equipped with the tool of a possible intervention, Woodward is able to define several causal notions: contributing direct and indirect causes, total causes, actual causes. Although Woodward does not pay any particular attention to explicating probabilistic causation, he does also account for probabilistic causation in his definitions of (in)direct contributing causation by postulating that 'a possible intervention on $X$ that will change $Y$ (or *the probability distribution of Y*)' (Woodward 2003: 59, emphasis added). This is an important point because at least some paradigmatic causal claims in the special sciences are about probabilistic causation, such as 'commercial hacienda systems *tend to* agrarian revolt' (cf. Kincaid 1996).

The next section provides information about the formal background of interventionist theories (i.e. about causal graphs and Bayesian networks). The main reason for introducing this formal background is to disentangle methodological and semantic questions concerning causation. The following section examines whether the interventionist theory is an adequate theory of causation for the special sciences. The success of the interventionist theory has to be judged by whether the theory satisfies the criteria of adequacy presented earlier: the naturalist criterion and the distinction criterion. The final section draws the conclusion that the interventionist approach is, overall, a promising candidate for an adequate explication of causation in the special sciences.

## The formal background of interventionist theories: causal graphs and Bayesian networks

As mentioned, interventionists aim at explicating causal notions that are taken as primitive in the interdisciplinary research on causal inference and causal modelling. Interventionists use causal models to explicate the causal notions in which they are interested, such as the notion of a direct type-level cause. Thus, the language in which the explications of causal notions are provided is (at least, partially) recruited from causal modelling. Let us consider some important notions in the framework of causal models.

The current debate on interventionist theories of causality is inspired by the actual practice and the (normative) methodology of causal modelling in econometrics, artificial intelligence and the medical sciences. In these contexts, variables are part of a *causal model* or a causally interpreted *Bayesian network*. Causal models and Bayesian networks are formal tools used to describe, to detect and to infer causal relationships.

A large part of the terminology – not merely the use of notions such as variable and value of a variable – used by interventionists to analyse the concepts of causation is drawn from these formal tools of modelling causality. A causal model is an ordered pair $\langle V, L \rangle$ of a set of variables $V$ and a set of equations $L$ specifying the relations among the variables in $V$. These equations take the form $Y = f(X, U)$, with $X, Y$ as variables that are members of $V$ and $U$ as an error term representing deviations and disturbances (cf. Pearl 2000: 27–9 see pp. 44–5). If the equations in $L$ are deterministic (which can be represented by an error term taking the value 0), the causal model is called a deterministic causal model – for simplicity's sake determinism is often assumed in the debate concerning the truth conditions of causal claims. If the equations are indeterministic, the resulting model is an indeterministic causal model. Indeterministic and deterministic causal models are alternatively referred to as causally interpreted Bayesian networks. We may view the deterministic case as a special case of a Bayesian network. Therefore, let us explore some more detailed information on Bayesian networks that is often implicitly used by proponents of interventionist theories of causation.

A Bayesian network[6] is an ordered pair $\langle G, Pr \rangle$, with a directed acyclic graph (DAG) $G$ consisting of and defined for:

- a set of variables $V$; and
- a set of directed edges connecting the variables, and a joint probability distribution $Pr$ over the variables of the graph.

The next sub-section introduces the components of a causal graph (i.e. variables and edges), and a probability distribution over a causal graph. This is followed by a presentation of the idea of structural equations that express the causal information contained in a causal graph more precisely than the directed edges in a DAG. Most importantly, this short introduction to causal Bayesian networks concludes by disclaiming two issues:

- although there are different rival methods of discovering causes by using the tools of Bayesian networks, these methodological questions concerning causal inference can be ignored when one is pursuing the semantic project of explicating causal notions; and
- debates about the success of the formal constraints on causal models (e.g. the causal Markov condition, the faithfulness condition, and the minimality condition) will not be addressed, as they are part of a separate (or, at least, separa*ble*) debate.

## A causal graph and its essential components: variables and edges

A graph consists of a set of variables V and a set of edges. If the edges are directed and interpreted as causal dependencies, the result is a causal graph. As previously mentioned, the notion of a variable plays a crucial role in the interventionist theory of causation. An intervention is defined as an intervention on a variable[7] – an intervention consists in the fact that an intervention variable takes a possible value such that the value of the cause variable changes (under certain restricted conditions). Let us briefly recapitulate the crucial information about variables for our purposes. In the literature on causation and probability theory, these variables are often called 'random variables'. According to the textbook understanding, a variable is a function from individuals (or outcomes) to the set of rational numbers: a variable $X$ is a function $X: D \rightarrow ran(X)$, with a set D of outcomes, and the set of possible values $ran(X)$ of the variable $X$. Assigning a value (which is a member of the set of possible values associated with a variable) to a variable, we may write a basic event statement of the form $X = x$ or $Y = y$. Statements of this form can be used to express information about quantitative event-types, qualitative event-types, and actual event-tokens of a quantitative or qualitative event-type. A basic event-type statement asserts that a variable takes one of its possible values. Basic event statements can be used to build complex statements by linking basic statements via logical connectives, such as the following complex statement $X = x$ *and* $Y = y$, *not-Y* $= y$, $X = x$ *or* $Y = z$, *if* $X = x$, *then* $Y = y$, *if* $X = x$ *and not-Y* $= y$, *then* $Z = z$ and so on. What is the content of an event statement such as $X = x$? An apt way to understand what variables represent is to think of them as representing a determinable property or a quantity. For example, a variable might stand for the determinable property of being coloured, and the values of this variable represent the determinants of the determinable such as being red, being blue, being green. Or consider a binary case: our variable $X$ represents throwing a ball. In this simplest case, this variable can take one of only two values: the ball is thrown ($X = 1$) or it is not thrown ($X = 0$). In many cases, it is even easy to translate the qualitative talk about events (and not event-types) into the talk about numbers: whether a system actually has the property of being F or not can be expressed by the values 1 (i.e. the system has the property F) and 0 (i.e. the system does not have the property F).

In the causal modelling framework, variables can be discrete or continuous.[8] The values of discrete and continuous variables are required to be mutually exclusive and exhaustive. The values associated with a variable $X$ are exclusive, if $X$ can only have one value (at a time). They are exhaustive, if the variable has to take one of its possible values. Suppose that Mary's income in September 2008 is either high or low; that is, the set of values associated with the income variable *I* is {high; low}. The set of values is exclusive, because in September 2008 Mary cannot have an income that is both high and low. The set is exhaustive if – as we suppose in the example – Mary's income has to be either high or low, and there is no further value that it might take – for example, 'medium' income.

Let me now turn to the remaining ingredients of a graph. I will introduce edges and arrows. In a graph the variables in **V** correspond to nodes, which are connected by edges. For example, variables *X* and *Y* are members of **V** and connected by an (undirected) edge (Figure 2.3). In the causal modelling literature, undirected edges represent correlations.

A directed edge, also called an 'arrow', gives more detailed information about the relationship between nodes (Figure 2.4).

A directed edge is the shortest directed path in a graph.

Although we are interested in causal graphs, there is no formal reason to interpret directed edges causally. They may also model logical relations (e.g. the relation '*Y* can be derived from *X*') or epistemic relations (e.g. '*X* is a reason for *Y*', '*X* confirms *Y*'). But, since we are interested in causality, directed edges are meant to represent the causal influence or dependence between the values of variables. According to this interpretation, the arrow introduced in Figure 2.4, connecting *X* and *Y*, means '*X* directly causes *Y*' or '*Y* directly causally depends on *X*'.

If, in a sequence of arrows, all edges point in the same direction, then this is called a *directed path* connecting more than two variables or a *chain* leading from one variable *X* to another variable *Y* via further intermediary variable *Z*. For example, take three variables *X*, *Y* and *Z* (see Figure 2.5).

If X is connected to *Y* by just one arrow, such as $X \rightarrow Y$ (here, $\rightarrow$ is a directed edge), then *X* is represented in the graph as the *direct cause* of *Y*. If *X* is connected to *Y* by a directed path comprising more than one arrow, such as $X \rightarrow Z \rightarrow Y$, then *X* is represented in the graph as an *indirect cause* of *Y*. There is also a convention to apply a 'family notation' to talk about the relation of nodes in a graph. For some *X* that is a member of

*Figure 2.3*   An undirected edge

*Figure 2.4*   A directed edge (i.e. an arrow)

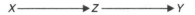

*Figure 2.5* A directed path (i.e. a causal chain, from $X$ to $Y$ through $Z$)

V in graph G, we can classify the direct and indirect causal relation by employing family notation.

- The set $Par_X$ of *parents* is the set of variables which are the direct causes of $X$; that is, the set of variables from which $X$ can be reached by one arrow and the arrow is directed towards $X$;
- The set $Chi_X$ of *children* of $X$ is the set of variables whose direct cause is $X$; that is, the set of variables which can be reached from $X$ by one arrow and the arrow is directed from $X$ towards the variables in this set;
- The *ancestors* of $X$ is the set $Anc_X$ which contains the parents of $X$, their parents, and so on;
- In turn, the *descendants* of $X$ consist of the set $Des_X$ whose members are the children of $X$, their children and so on.

Family notation is a useful metaphor, often used to simplify the formulation of formal conditions that are supposed to hold for a graph, such as the causal Markov condition (see p. 47).

In the standard usage of causal graphs it is assumed that the graph contains no causal loops. A causal loop is a directed path in which the same variable appears more than once, such as $X{\to}Z{\to}Y{\to}X$. A graph of this structure is known as *cyclic*. An *acyclic* graph that has only directed edges is called a *directed acyclic graph* (DAG). A DAG is the standard[9] graphical representation of a causal relation in the current philosophical debate about causation.

In the literature on causal modelling, the set of variables V is usually split into two subsets: a set of *endogenous* variables E and a set of *exogenous* variables U; that is, the set V is the union set of all endogenous and exogenous variables. Exogenous variables are taken as given in the model, because their values are not described as being caused by other events. They have no causal predecessors in the graph. Rather, the role they play is restricted to being the causes of endogenous variables. In contrast, each endogenous variable does have a cause in the graph. Note that there is a clear distinction between exogenous and endogenous variables *within* a given graph. But this is not a principled distinction: one variable might be considered as endogenous in one graph and as exogenous in another graph. Whether a variable is endogenous or not depends mostly on the pragmatic reasons of modelling scientists.

As introduced above, a Bayesian network is an ordered pair of $\langle G, Pr \rangle$. A joint probability distribution over a graph, usually a causally interpreted DAG, composes a Bayesian network. But what is meant by a probability distribution? The probability distribution specifies the meaning of the edges. More precisely, it specifies the relation between some variable $X$ and its parents. It does so by providing a conditional probability function: it specifies the probability of each value of $\{x_1, ..., x_n\}$ of $X$ conditional on each value of $\{par_{X1}, ..., par_{Xn}\}$ of the parents$_X$.[10]

### Structural equations

A Bayesian network is supposed to provide causal information. Some philosophers and scientists assume explicitly that this information can be expressed in a specific mathematical form: it is mathematically expressed in so-called *structural equations*.[11] Structural equations provide the most precise description of the relations in a graph: the equations represent how a variable depends on another. Structural equations describe these relations among variables with greater precision than the directed arrows in a DAG. A structural equation determines the value of a dependent variable $Y$, the effect, given a (set of) endogenous variable(s) $X$, the direct causes or causal parents of $Y$, and the exogenous variables $U_i$. In economics and the social sciences, structural equations usually take the following linear form: $Y = \alpha X + \beta U$, with $\alpha$, $\beta$ as parameters representing the strength of influence that $X$ and $U$ have on $Y$ (cf. Cartwright 1989: ch. 4 s. 5, Pearl 2000: 27 and ch. 5, Hoover 2001: ch. 2). More neutrally and with less commitment to the linear character of structural equations it may be written as a function: $y = f_Y(x_i, u)$ with the $X_i$ as the set $par_Y$ of parents of $Y$. By convention, we may call a causally interpreted Bayesian network a 'structural causal model', if its causal information can be expressed by a set of structural equations. Note that the algebraic symbol '=' is given an asymmetric interpretation: in the case of structural equations 'side matters'.[12] Therefore, '$Y = f_Y(X_i, U)$' is not equivalent to '$f_Y(X_i, U)$' = $Y$. By convention, the independent variables (namely, the causes) are written on the right-hand side of the equation, while the dependent variables (namely, the effect) appear on the left-hand side.

Further, interventionists take structural equations to 'encode' counterfactual information. The structural equation $Y = f(X, U)$ encodes the counterfactual information thus: for example, if it were the case that $X = x$ and $U = u$, then it would be the case that $Y = y$. The value of the dependent variable $Y$ is a function of the values of $X$ and $U$; that is, the value of $Y$ is assigned according to the function f in the structural equation for counterfactual values of $X$ and $U$ (Hitchcock 2001: 280). Interventionist theories of causation commonly adopt the use of structural equations in order to describe the relation between cause and effect, and to describe what will happen under interventions (particularly, structural equations are taken to be a way of expressing that the relations among variables will stay invariant in the case of an intervention; invariant generalizations are discussed in Chapter 5).

## Formal conditions of causal inference and disclaimers

To complete the presentation of Bayesian networks further essential formal constraints should be noted, such as the causal Markov condition (p. 47). However, simply stating the nature of a Bayesian network does not explain the epistemic value of these formal tools. The very reason why many scientists and philosophers are enthusiastic about Bayesian networks lies in its epistemic virtues: Bayesian networks are powerful tools with which to discover causal relationships and, in particular, to infer causes from acausal statistical data. It is a majority opinion in the debate that causal models can be inductively inferred from statistical data (cf. Pearl 2000, Spirtes et al. 2000). However, the method of inductive causal inference is by far not the only one to have been proposed in order to detect causes by means of Bayesian networks. It is also a matter of disagreement whether causal models can be confirmed (let alone be inductively inferred) from purely statistical data without using any prior causal information (cf. Cartwright 2007). Rather, there are a number of different methodological approaches to the task of testing causal statements in the framework of Bayesian networks. The three most important methodologies in the literature are the method of inductive learning (causal inference), the hypothetico-deductive method, and the hybrid method of discovery. Let me briefly sketch the basic idea behind these three methods (cf. Williamson 2005).

*Inductive learning*: Bayesian network theorists such as Spirtes et al. and Pearl advocate inductive rules of inference (i.e. algorithms) that are supposed to enable researchers to infer causal models directly from non-causal, statistical data. The causal models that result from using these algorithms are supposed to have the structure of a DAG that satisfies the causal Markov condition, the minimality condition, and the faithfulness condition). Most prominently, Spirtes et al. (2000: s. 5.4.2) advocate the PC algorithm (named after Peter Spirtes and Clark Glymour) (cf. Williamson 2005: 123, for a less technical exposition of the algorithm). Pearl (2000: s. 2.5) adds further conditions to the PC algorithm and also develops another algorithm, the inductive causation (IC) algorithm.

*Hypothetico-deductive method*:[13] Using this method, a causal model is not tested by inferring causes directly from statistical data. Rather, the model is tested by the following three-step methodological procedure:

1. Assume that there is a causal relationship; that is, hypothesize this causal relationship, and express this assumption in a general statement (a law) that is part of a Bayesian network (e.g. in the form of a structural equation);
2. Deduce the consequences from the assumed general statement;
3. Return to step 1, if the assumption contradicts evidence.

*Hybrid method of discovery* (Williamson 2005: 148): This method follows a four-stage rule that combines elements of the hypothetico-deductive approach (on the first and second stage) and the inductive learning approach (on the third and fourth stage):

1. Hypothesize: A causal graph is induced from constraints;[14]
2. Predict: Predictions are deduced from the hypothesized graph;
3. Test: Evidence is obtained to confirm or disconfirm the hypothesis;
4. Update: The causal graph is updated in the light of new evidence.

In the debate on causal modelling and causal inference, the best method by which to discover causality is a matter of controversy.[15] Even though I would not wish to deny that resolving this controversy is an interesting philosophical (and even interdisciplinary) project, I remain convinced that the epistemic questions of causal inference – and, more generally speaking, the question of how empirical testing of causal models works – can, by and large, be disentangled from the semantic project with which interventionist theories of causation deal. Permit me explain this claim.

Regardless of the question as to which of the three methods is to be preferred, advocates of a Bayesian network approach to testing causal statements seem to agree on the formal constraints that have to be imposed on a causal model. The two components of a causal Bayesian network, the causal graph and the probability function over the graph, are constrained by several formal requirements. So, before we can turn to the important point of this section – the disclaimers – it is useful to understand the formal constraints to which a Bayesian network, or a causal model, has to conform. There are three important formal constraints, although this list of constraints is not exhaustive.

The most fundamental formal requirement is the Markov condition.

## Markov condition

Conditional on the parents of a variable $X$ (i.e. the set of variables that can reach $X$ by one arrow which is directed towards $X$), a variable $X$ is probabilistically independent of its non-descendents and its non-parents (cf. Spirtes, Glymour and Scheiner 2000: 11).[16]

*Probabilistic independence* is defined as follows: $X$ is probabilistically independent of Y iff the probability that $X$ takes one of its values conditional on Y *is equal to* the probability that $X$ takes one of its values; that is, $P(X = x_i | Y = y_j) = P(X = x_i)$. In the case of the Markov condition, given the parents, all other ancestors are probabilistically irrelevant for the assignment of a value to $X$. Nevertheless, $X$ is not independent of its descendents (i.e. those nodes in the graph, or variable, which can be reached from $X$ by one or more directed arrows).

A *causally interpreted* Bayesian network is constrained by the causal interpretation of the Markov condition. As stated, a Bayesian network satisfying the Markov condition does not have to be interpreted to represent causes. It could alternatively be used to represent confirmation relations. However, we are interested in Bayesian networks as representations of causal relations. The causal interpretation of the Markov condition is defined as follows:

## Causal Markov condition

Conditional on its direct causes (i.e. in terms of the family notation: the set of its causal parents), a variable $X$ is probabilistically independent of its non-effects (cf. Pearl 2000: 30, Spirtes, Glymour and Scheiner 2000: 29f.).

One can motivate the value of the causal Markov condition by showing that one can derive plausible theorems therefrom (cf. Schurz 2006: 242f.).[17] One of these theorems is a well-known assumption in philosophy of science: the *Common Cause Principle* that was introduced in the philosophical debate by Hans Reichenbach (cf. Reichenbach 1956; Suppes 1970). The common cause principle connects correlations and causation (i.e., to be precise, possible causal relations).

## Reichenbach's common cause principle

If two variables $X$ and $Y$ are probabilistically correlated, then this correlation indicates that one of the three following possible causal relations must hold:

- $X$ causes $Y$,
- $Y$ causes $X$,

- there is some set of variables $\{C_1, ..., C_n\}$ which is the set of common causes of $X$ and $Y$ (or, of common causal predecessors),[18] such that $X$ and $Y$ are probabilistically independent conditional on $\{C_1, ..., C_n\}$.

Reichenbach famously also uses the concept of *screening off* in order to express probabilistic independence, such as '$X$ and $Y$ are *screened off* by their common causes $\{C_1, ..., C_n\}$'. Schurz (2006: 242f.) explains and motivates the value of the causal Markov condition by deriving two further theorems therefrom.[19] The second theorem states: if neither $X$ causes $Y$ nor $Y$ causes $X$, then $X$ and $Y$ are probabilistically independent conditional on all the common causes $\{C_1, ..., C_n\}$ of $X$ and $Y$. The third theorem that can be derived from the causal Markov condition requires, according to Schurz, that $X$ is probabilistically independent of its indirect causes conditional on the set of direct causes of $X$; likewise, $Y$ is probabilistically independent of its indirect causes conditional on the set of direct causes of $Y$. These three theorems can be taken to motivate the causal Markov condition, because the theorems are, indeed, plausible.

The causal Markov condition is a fundamental constraint for Bayesian networks. (It is also a strong assumption, and many critics of Bayesian networks as a formal tool for modelling causal relationships claim that the condition does not hold for the description and explanation of the causal interactions of all systems.) Besides the causal Markov condition, there are further conditions that are taken to be essential formal constraints on a causally interpreted Bayesian network. The two most important are the minimality condition and the faithfulness condition. The fundamental character of the causal Markov condition is partly revealed by the fact that the definition of both conditions uses the causal Markov condition in the *definiens*. Note that the following formal conditions are methodological assumptions made by modelling scientists in order to infer causal models from purely statistical data sets. Let us first focus on the minimality condition.

### Minimality condition

A Bayesian network $\langle G, Pr \rangle$ is minimal iff no proper subgraph of the graph G satisfies the causal Markov condition (cf. Pearl 2000: 46; Spirtes et al. 2000: 12).

The minimality condition is supposed to guarantee that a probabilistic dependence relation corresponds to every arrow in the graph. More loosely speaking, the minimality condition is a formal expression of the simple idea that the representation of causal relations (here, a Bayesian network) should contain only relevant information and leave out irrelevant information. Additionally, a Bayesian network is supposed to fulfil the faithfulness condition:

**Faithfulness condition**
The probability distribution on a Bayesian network $\langle G, Pr \rangle$ is faithful iff the probability distribution contains all and only those conditional probabilistic independence relations which are entailed by the causal Markov condition (cf. Pearl 2000: 48; Spirtes et al. 2000: 13).

The faithfulness condition is most clearly a methodological assumption: it is required to hold by modellers in order to infer causal independence relations from probabilistic independence relations. It can be interpreted to describe a norm, which leads us to prefer models that do not rely on hidden causal relations between variables. The minimality condition can be understood as a methodological tool for eliminating redundant information. For this reason, the two conditions are sometimes called the 'formal version of Occam's Razor' (cf. Pearl 2000: 45f.).

This section presented methodological accounts of how to test causal statements in the framework of Bayesian networks (the inductive, the hypothetico-deductive, and the hybrid method). It then outlined three important formal constraints that are essentially imposed on causally interpreted Bayesian networks (the causal Markov condition, the minimality condition, and the faithfulness condition). Equipped with this background knowledge concerning the methodological debate, I now present two disclaimers.

*Disclaimer 1:* The inquiry in this book concerns the conceptual analysis of causation and accounts of truth conditions of causal claims; that is, I engage in the semantic project with respect to causation (see Chapter 1, pp. 3–5). For this reason, the question as to which methodology turns out to be the right one for inferring causes from statistical data is somewhat of a side issue. This view is supported by the way in which Woodward demarcates his central semantic goal 'to give an account of the content or meaning of various locutions, such as $X$ causes $Y$' (Woodward 2003: 38) from methodological projects. Contrasting his approach with the methodological projects of Pearl (2000) and Spirtes et al. (2000), Woodward declares: 'By contrast, I have nothing to say about issues having to do with calculating quantitative magnitudes, estimation, identifiability, or causal inference' (Woodward 2003: 38). Woodward asserts that the work of Pearl, and Spirtes et al. mainly focuses on those methodological tasks to which he refers in the quote. Woodward continues in the next sentence to outline his own enterprise as being a part of the semantic project with respect to causation:

> Instead, my enterprise is, roughly, to provide an account of the meaning or content of just those qualitative causal notions that Pearl (and perhaps Spirtes et al.) take as primitive. [...] *my project is semantic or interpretative, and is not intended as a contribution to practical problems of causal inference.* (Woodward 2003: 38, emphasis added)

In other words, Woodward emphasizes that his semantic project and the methodological projects are sufficiently separate enterprises. I do not wish to claim that the methodological and the semantic project with respect to causation do not have any mutual impact on each other: quite the contrary. For instance, if one provides truth conditions for causal claims without the vaguest hint as to how one might acquire knowledge of whether a particular causal statement is true, then the semantic project is philosophically not fruitful. However, it is unlikely that this is the case for interventionist theories, because the interventionist theory of causation can be (and, in fact, is) easily reconciled with various accounts of testing causal claims. It is, to say the least, not beyond comprehension to know how to test a causal claim whose truth conditions are stated in the interventionist framework. Moreover, one can remain optimistic by pointing out that there is, fortunately, a division of cognitive labour between semantic and methodological theories of causation (see Chapter 9, p. 250).

The separation of semantic and methodological questions also has another interesting consequence: it precludes interpreting Woodward's theory of the meaning of causal statements along verificationist lines. Prima facie, one could ascribe a verificationist theory of meaning to Woodward: the meaning of a causal claim '$X$ causes $Y$' consists in the conditions that have to obtain in order to test the claim (testability conditions). In the case of Woodward's interventionist theory, these testability conditions amount to a hypothetical experiment that consists in carrying out an intervention on $X$ (recall, for instance, the randomized controlled experiment: pp. 29–31). However, there are good reasons to reject the verificationist reading of Woodward's semantic project: first, Woodward clearly separates semantic and methodological questions. Second, the verificationist reading conflicts with Woodward's remarks about the meaning of causal statements. Woodward insists that the meaning of causal statements is crucially related to truth conditions (rather than testability conditions) and objective, mind-independent truth-makers (cf. Woodward 2003: 118–23).

*Disclaimer 2*: There is extensive debate about the validity of the formal requirements (in particular, the causal Markov condition and the faithfulness condition are discussed) that causal models should satisfy (among the critics are Gillies 2001; Steel 2005, 2006; Cartwright 2007). Since interventionists use the framework of causal modelling in order to explicate causal notions, it would be a problem for their approach if it transpired that these crucial assumptions did not hold. However, this is a different debate into which this book cannot enter. Some justification for ignoring this debate can be found in the fact that defenders of causal models and Bayesian networks seem to make a convincing case in defence of their project, in general, and the formal constraints that are imposed on causal models, in particular (cf. Hausman and Woodward 1999, 2004; Glymour 2010). To say the least, it would seem that the controversy about the formal requirement on causal models has not yet been decided in disfavour of the advocates of Bayesian networks.

## Is the interventionist theory adequate?

Why should we adopt the interventionist theory of causation? Is the interventionist theory an adequate account of causation for the social sciences in particular? Is it an adequate account of causation for the special sciences in general? Earlier in this chapter, it was claimed that the interventionist theory of causation is acceptable as an adequate explication of causation in the context of these disciplines only on the condition that it satisfies the naturalist criterion and the distinction criterion. This chapter will investigate whether Woodward's interventionist theory conforms to the naturalist criterion (i.e. characteristic features and kinds of causation in the special sciences). The result of this investigation will be mainly positive (although there are several desiderata). The chapter will then turn to the distinction criterion of an adequate explication of causation. It will be shown that interventionist theories are able to produce a correct and natural description of various intuitively different causal structures that trouble many other theories of causation as counter-examples. In particular, the chapter will discuss several scenarios of pre-emption and the controversial assumption that causation is a transitive relation. The chapter concludes with the observation that interventionist theories of causation promise to be good candidates for an adequate explication of causation in the special sciences.

## The naturalist criterion of adequacy

Let us now go through the list of typical kinds and characterizing features of causation (Chapter 1, pp. 15–20) step by step and consider whether the interventionist theory can capture the feature of causation in question.

1. *Type-level and token-level causation*: By and large, the special sciences are interested in type-level causes as well as actual token-causes. Type-level causation and actual causation differ with respect to the relata standing in the causal relations: type-level causation relates event-types and actual causation relates event-tokens.

Interventionists are able to define various notions of type-level causation (e.g. direct, indirect, total, and contributing causation) (pp. 26–39). The notion of actual causation can also be accounted for. It is explicated derivatively of type-level causation in the sense that event-types (represented by statements of the canonical form $X = x$) are taken to be instantiated by actual objects. That is, statements of the canonical form $X = x$ are considered to be actually true when describing token-events as the relata of actual causation. In sum, the interventionist theory captures this feature of causation, because the theory provides definitions for various kinds of type-level causation, and it also defines actual causation.

2. *Deterministic and probabilistic causation*: According to the paradigmatic cases of causal statements in the social sciences, some causal relations are deterministic, while other causal relations are probabilistic (i.e. the causes make a probabilistic difference to the effect).

Although interventionists more frequently refer to deterministic causation, probabilistic causation is also addressed in the definition of contributing causes. Woodward does account for probabilistic causation in his definitions of (in)direct contributing causation by postulating that 'there be a possible intervention on X that will change Y (or *the probability distribution of Y*)' (Woodward 2003: 59, emphasis added). In other words, Woodward accounts for this kind of causation, because he provides definitions for various kinds of deterministic and, implicitly, indeterministic causation.

3. *Multiple causes and degrees of causal influence*: Social scientists hold that real-life cases of causation usually involve multiple causes of a phenomenon, and these multiple causes contribute to the effect in varying degrees.

From an interventionist point of view, the most plausible interpretation of 'multiple causes' is that a phenomenon has a number of contributing causes. For this reason, interventionists should have little trouble with this requirement, because nothing in the interventionist definition of contributing causes precludes that an effect might have several contributing causes. Interventionists are also able to capture the possibly varying degree of influence that several different causes may have on the same effect: invariant generalizations (and structural equations) can quantify the degree of influence in the form of coefficients (such as in the equation $Y = \alpha X + \beta Z$, with $\alpha$, $\beta$ as coefficients representing the degree of influence that $X$ and $Z$ have on $Y$). Thus, interventionists conform to this criterion of adequacy also.

4. *Modal relation*: Causes enforce, bring about and produce their effects. In other words, some kind of modal connection obtains among cause and effect.

Hitchcock (2007: 57f.) set the agenda for any philosopher who wants to participate in building a theory of causation. This agenda consists of two central demands Hitchcock calls 'the hard problems of causation'. One of the hard problems consists in explaining the 'modal force' of a causal relation in a way that distinguishes it from accidental correlations of events. Two claims must be distinguished here: first, there is a distinction in kind between genuine causation and accidental correlations; second, one can account for this distinction by specifying a 'modal force' that is attached to causal relations. Interventionists appear to adopt the first claim, but they remain silent with respect to the second.

How can we distinguish between accidental and causal relations? One way to draw this distinction is to claim that causation is a physical process or a physical interaction involving conserved quantities (cf. Dowe 2000). On this view, accidental correlations fail to meet this condition. Another way to draw the distinction between causation and accidental correlation is by noting that causation is a modal relation whereas accidental correlation is not. For instance, one might claim that causation is a conceptually and metaphysically primitive kind of production that is not able to be further explicated (cf. Machamer 2004; Moore 2009). Yet another modal option is to view causation as the instantiation of a necessitation relation (cf. Armstrong 1983). And, naturally, there are numerous alternative views of modal force. However, interventionists are not committed to any one of those metaphysical options to distinguish causes from accidents when they are defining type-level causation. They can be interpreted as giving a minimal version of the claim that one can account for this distinction by specifying a 'modal force': modal force could be viewed as the restricted necessity that is a feature of (interventionist) counterfactual conditionals (cf. Lewis 1973a: 13–19). In a sense, one could term this necessity as conditional nomological necessity, because it is grounded in the structural equations conditional on special initial conditions. Given that (a) an intervention on the cause variable occurs, (b) other off-path causes are held fixed in the redundancy range, and (c) given that certain deterministic structural equations hold, then the occurrence of the effect follows with necessity.[20] This nomological necessity does not imply any particular metaphysical account of the 'modal force' associated with causation. All of the metaphysical options (e.g., physical processes, primitive production, necessitation and so on) are still available. But why should this amount to a problem? Interventionists *can* capture the distinction of causal relations and accidental correlations by referring to interventionist counterfactuals and structural equations. The core of this distinction is the interventionist theory of laws: the *invariance theory*. Causal relations are described by invariant generalizations; accidental correlations fail to be invariant. In sum, interventionists provide a minimal account (the invariance theory of laws) for drawing the distinction between accidental correlation and causal relations.

The invariance theory of laws has to face several pressing questions: the relation between counterfactual conditionals and structural equations (or, more broadly speaking, invariant generalizations) is a delicate and difficult matter. Critical discussion of their relationship is offered in Chapters 3, 4, 5 and 7. Chapters 3 and 4 focus on the semantics for interventionist counterfactuals, Chapter 5 deals with the role of counterfactuals in defining the invariance of generalizations, and Chapter 7 discusses and defends conceptually circular explications of causal concepts.

5. *Time-asymmetry of causation*: Causes precede their effects in time, and not vice versa.

Interventionists claim that the time-asymmetry of causation can be explained by appealing to possible interventions. Recently, interventionists have presented an argument for this claim, which I call the 'open-systems argument' and which will be discussed in detail in Chapter 6. For now, let us take this as a promissory note to the effect that interventionists are able to account for the time-asymmetry of causation.

6. *Causal asymmetry*: The causal relation is asymmetric; that is, if $X$ causes $Y$, then $Y$ does not cause $X$.

Causation is also asymmetric in another sense: the causal relation is itself asymmetric, for if $X$ causes $Y$, then $Y$ does not cause $X$. Moreover, interventionists typically postulate that interventionist counterfactuals are non-backtracking. One can easily imagine that causation might be asymmetric in this latter sense and, even so, fail to be time-asymmetric. Thus, it seems that the causal asymmetry does not imply the time-asymmetry of causation, but time-asymmetry implies the causal asymmetry. The decisive question is whether interventionists can account for causal asymmetry. The answer to this question is not straightforward. Certainly, interventionists claim that the most promising strategy to explain causal asymmetry is to appeal to possible interventions. And, according to interventionism, it is true by definition that $X$ causes $Y$ iff there is a possible intervention on $X$ that changes $Y$. According to interventionists, the reverse does not hold: if $X$ causes $Y$, then there is no possible intervention on $Y$ that changes $X$ as well. In other words, the corresponding interventionist counterfactuals are taken to be non-backtracking. However, stipulating that causation is asymmetric does not amount to explaining why this asymmetry holds. To explain the asymmetry is what we would expect of an adequate theory of causation. Thus, the interventionist theory does not capture the feature of causal asymmetry in a straightforward manner. Yet, one can attribute an indirect explanation of the causal asymmetry to interventionists: if it is true that the time-asymmetry of causation implies the causal asymmetry and if interventionists provide a sound argument for the time-asymmetry of causation (the open-systems argument), then interventionists have an indirect explanation of causal asymmetry. However, the success of explaining the time-asymmetry of causation depends on the soundness of the open-systems argument, which will be discussed in Chapter 6.

7. *Distinction of spurious correlation and genuine causation*: There is an important distinction between (a) the case of a correlation between $X$ and $Y$ (maybe due to a common cause $Z$), and (b) the case of a causal relation between $X$ and $Y$ (i.e. either $X$ causes $Y$, or vice versa).

A classic example illustrates this criterion. Suppose that (a) barometer reading 'storm' and the occurrence of storms are correlated, (b) it is neither the case that the barometer reading 'storm' causes the occurrence of storms, nor is it the case that the occurrence of storms causes the barometer reading 'storm'. What is the case is that (c) the barometer reading 'storm' and the occurrence of storms are the effects of a common cause; namely, the drop in atmospheric pressure. The upshot of the example is that barometer readings and storms are merely correlated, while there is a genuinely causal relation between the atmospheric pressure and the correlated event-types. Are interventionists able to distinguish mere correlation between barometer readings and storms from a causal relation between barometer readings and storms? I think they clearly can: it is neither the case that an intervention on the barometer reading changes the occurrence of a storm, nor that an intervention on the occurrence of a storm changes the barometer reading. Therefore, according to the interventionist approach, it is not the case that the barometer reading 'storm' causes the occurrence of storms, neither is it the case that the occurrence of storms causes the barometer to read 'storm'. However, if there is an intervention on the common cause – the atmospheric pressure – then the barometer readings and the occurrences of storms change correspondingly. In other words, genuinely causal relations are described by invariant generalizations, while mere correlations are not. Thus, interventionists are able to distinguish spurious correlations from genuine causation.

8. *Context*: Many of the examples given here of causal statements are (implicitly) believed to hold only on the assumption that other variables are held fixed. This assumption is often referred to (by economists) by the expression 'ceteris paribus'.

The fact that context matters, in the sense that causal statements are believed to be true only on the assumption that other variables are held fixed, is explicitly built into the interventionist definitions of causal notions. Contributing type-level causes and actual causes are defined with respect to a redundancy range. Analogously, an intervention variable is, by definition, required not to alter the relationship between $Y$ and its causes $W_1, ..., W_n$, which are not on a causal path leading from $X$ to $Y$. Moreover, the holding-fixed of the causal context is also essentially employed for the purpose of characterizing the invariance of structural equations, which describe what happens if an intervention on a variable occurs (the invariance of generalizations is discussed in Chapter 5). In sum, the interventionist approach performs extremely well relative to the context-criterion of adequacy.

9. *No-universal-laws requirement*: Since there are no universal laws in the social sciences (and, possibly, in the special sciences in general), an explication of causation for these disciplines cannot refer to such laws.

Does the interventionist theory satisfy this naturalist criterion of adequacy? It does, because it does not require universal laws in order to state the truth conditions of causal statements. Although interventionists do not appeal to universal laws, they do refer to a weaker kind of generalization when they claim that the relationship between variables remains invariant under interventions. These generalizations do not satisfy the characteristics of laws as philosophers traditionally conceive them. Although they lack several 'standard' features of lawhood, (minimally) invariant generalization can be distinguished from merely accidentally true, universal statements. These important issues are covered in Chapter 5. For the time being, we may conclude that, prima facie, interventionist theories of causation satisfy the no-universal-laws requirement.

Let us briefly take stock. The interventionist theory of causation has been confronted with nine criteria of adequacy which are subsumed under the naturalist criterion for the explication of causation in the special sciences. Prima facie, the interventionist approach conforms to all of the nine sub-criteria of the naturalist criterion. However, whether or not interventionists are able to cope with at least three of these criteria (i.e. on the one hand, the no-universal-laws requirement, and, on the other hand, the time-asymmetry and the asymmetry of causation) will have to be the subject of a more detailed discussion in Chapters 5 and 6. For now, it does not seem to be too hasty to declare that the interventionist theory of causation conforms to the naturalist criterion in a promising way.

## The distinction criterion of adequacy

According to the distinction criterion, the explicated causal concept has to be able to correctly distinguish between intuitively different causal structures. Observe that one of the criteria belonging to the naturalist criterion, the distinction of spurious correlation and genuine causation, is in fact itself a distinction criterion that social scientists hold in high regard. The general motivation to appeal to the distinction criterion consists in the idea that the application of concepts should be tested not only for actual cases, but also for those merely possible cases we would classify as cases in which the concept should apply. It will be shown that interventionist theories are able to give correct and natural descriptions of causal structures that trouble many other theories of causation as counter-examples. In particular, discussion will include several scenarios of pre-emption and the controversial assumption that causation is a transitive relation.

The main reason to focus on pre-emption and transitivity consists in the fact that both have proven to be tenacious counter-examples to the most influential theory of causation in most areas of philosophy: the counterfactual theory of causation. The counterfactual theory was originally formulated by David Lewis (1973b). Lewis's main target is actual causation; that is, causation between event-tokens (cf. Lewis 1973b: 161f.). Lewis defines actual causation in two steps: he first defines causal dependence, and then proceeds to define causation in terms of causal dependence. According to Lewis (1973b: 165–7), for two distinct possible events $c$ and $e$, $e$ causally depends on $c$ iff the following counterfactuals are true: if $c$ were to occur, then $e$ would occur; and if $c$ were not to occur, then $e$ would not occur. According to Lewis's possible worlds semantics for counterfactual conditionals (cf. Lewis 1973a), a counterfactual is true iff, roughly, the consequent of the conditional is true at all of the closest possible worlds in which the antecedent of the conditional is true. Lewis's semantics for counterfactuals will be elaborated in Chapter 3. Let us return to Lewis's analysis of causation. Lewis (1973b: 167) defines causation as a transitive relation: be $c$, $d$, $e$ ... a finite sequence of distinct events such that $d$ causally depends on $c$, $e$ causally depends on $d$ and so on: a causal chain. Lewis defines actual causation thus: 'one event is the cause of another iff there exists a causal chain leading from the first to the second' (Lewis 1973b: 167).

In his recent *Whitehead Lectures*, Lewis[21] revises and specifies his counterfactual account of causation by analysing causation in terms of causal influence, which states:

> C influences E iff there is a substantial range $C_1$, $C_2$, ... of different *not-too-distant alterations* of C (including the actual alteration of C) and there is a range $E_1$, $E_2$, ... of alterations of E, at least some of which differ, such that if $C_1$ had occurred, $E_1$ would have occurred, and if $C_2$ had occurred, $E_2$ would have occurred, and so on. Thus we have a pattern of counterfactual *dependence of whether, when, and how* [E occurs – AR] *on whether, when and how* [C occurs – AR]. [...] [C]ausation is the ancestral: C causes E iff there is a chain of stepwise causal influence from C to E. (Lewis 2004: 91, emphasis added)

Lewis's new model of causation can be understood as a quantitative revision of his classic, qualitative approach. In the quantitative approach, the causal relata (the events) have become more fine-grained. Lewis claims that an event can occur differently in many respects. One can draw an illustrative analogy between Lewis and Woodward: Lewis, as well as adherents of the interventionist theory, wants to make room for the idea that cause and effect cannot simply *only occur* or *not occur* – they can occur in many different ways. Lewis spells out this idea by introducing 'different *not-to-distant alterations* of C', while interventionists can model actual causes and effects as the actual instantiation of a quantitative event-type $X = x$ (where $x$ is the possible value of a quantitative variable $X$). However, where exactly the differences between Lewis's original theory and its revised version lie is not important for our purposes. More important, in the context of discussing the distinction criterion of adequacy, is to note two crucial features that Lewis's original theories and his revised theories of causation have in common.

First, Lewis's theory of causation analyses causation in terms of counterfactuals. Lewis argues that analysing causation in terms of counterfactuals (and causal dependence) leads to the view that causes are necessary for the occurrence of their effects. Scenarios of pre-emption are a threat to the counterfactual theory, because they challenge precisely this latter claim that causes are necessary for the occurrence of the effect. In scenarios of pre-emption, each cause is not necessary for the effect since other pre-empted causes are sufficient to bring about the effect. This kind of counter-example to Lewis's theory will be explained in greater detail (pp. 60–7).

Second, Lewiss analysis defines causation as a transitive relation. However, the assumption that causation is a transitive relation faces counter-examples. One of these counter-examples against Lewis's counterfactual theory will be discussed (pp. 66–9).

Since Woodward's interventionist theory of causation is also a version of a counterfactual theory of causation – because it relies on interventionist counterfactuals to analyse causal notions – I will examine whether Woodward's theory can describe scenarios of pre-emption correctly, and whether interventionists can reject counter-examples to the assumption that causation is a transitive relation. Let me emphasize that I will restrict my attention to the success or failure of interventionist theories. I will not discuss whether Lewis's original theory of causation or his revised version of the theory can be saved from all scenarios of pre-emption and from counter-examples to the claim that causation is transitive. However, many philosophers doubt that Lewis's approach can be vindicated on all fronts (cf., for instance, discussions in Collins et al. 2004a, and Price and Corry 2007).

Before turning to the causal scenarios, let me add a brief remark on whether Lewis's theory satisfies the naturalist criterion. An important observation is that Lewis merely provides an account of actual causation. Lewis does not provide an analysis of the variety of type-level causes (see pp. 15–16). This is certainly a shortcoming with respect to the naturalist criterion. Thus, Lewis's theory is not only troubled by counter-examples falling under the distinction criterion, his approach also fails to meet the naturalist criterion to a satisfying degree. One may defend Lewis on the grounds that his counterfactual theory of causation is merely intended to provide an analysis of the common-sense concept of causation, rather than the concept of causation as it is paradigmatically used in the special sciences. However, the objection remains the same: even if one focuses on the common-sense concept of causation, speakers in non-scientific everyday contexts refer not only to actual causation, but also to (deterministic and probabilistic) type-level causation. Thus, Lewis's theory seems to be inadequate in this respect, even if it aims to be an analysis of the common-sense concept of causation.

*Scenarios of pre-emption*

In pre-emption cases, the effect, $C$, is over-determined by at least two causes, $A$ and $B$, in the sense that each of the causes is sufficient for the effect. The causal graph in Figure 2.6 represents the general structure of pre-emption cases: $C$ is over-determined by $A$ and $B$.

As outlined, pre-emption cases are especially a threat to those theories of causation that endorse the claim that causes are necessary for their effects (such as Lewis's counterfactual theory of causation), because each cause (such as $A$) is *not* necessary for the effect since *other* potential but pre-empted causes (such as $B$) are *sufficient* to bring about the effect. In the literature, pre-emption is discussed in (at least) four ways: as *early* pre-emption, *late* pre-emption, *trumping* pre-emption, and *symmetric over-determination*. The various scenarios of pre-emption mainly differ with respect to the temporal relations among $A$, $B$, and $C$.

In the case of early pre-emption, the potential cause $B$ is interfered with before the actual causal process from $A$ to $C$ takes place. With late pre-emption, the pre-empted cause $B$ occurs shortly after the actual causal process from $A$ to $C$. In trumping pre-emption cases, $A$ and $B$ occur at the same time, yet $A$ always overrules or trumps $B$ as a cause of $C$. According to a scenario of symmetric over-determination, $A$ and $B$ occur at the same time, and neither $A$ trumps $B$, nor vice versa. One

*Figure 2.6*   General structure of a pre-emption scenario

seems to have equally good reasons to claim that both *A* and *B* are causes of *C*. Let us examine an example for each variant of pre-emption.

*Early pre-emption*: Let us imagine the following situation (cf. Lewis 1986a, 2004; Hall 2004): two assassins, A and B, have conspired to shoot the dictator of their country. They break into a building across the street from where the dictator is giving a speech. From the roof of the building, both take aim, side by side. Both are excellent marksmen and never miss their target. Assassin A fires a shot and the dictator is killed in the middle of his speech. B desisted from shooting his gun, because he saw that A was pulling the trigger, heard the shot and so on. Both escape successfully through the back door of the building. We may represent the situation with the following variables and their values (cf. Woodward 2003: 77–9, Halpern and Pearl 2005: 861f.).

- *Exogenous variables* $U_i$: $U_A$ represents the motivation of assassin A to kill, to plan the killing and to take actions that are necessary preparations for the assassination and so on; respectively, $U_B$ represents the same motivation for B. A being motivated to kill is represented by $U_A = 1$, A not being motivated is represented by $U_A = 0$. Analogously for $U_B$, $U_A$ and $U_B$ are supposed to take the value 1.
- *Endogenous variable A*: *A* represents the shot of assassin A, shooting is represented by $A = 1$, not shooting by $A = 0$.
- *Endogenous variable B*: *B* represents the shot of assassin B, shooting is represented by $B = 1$, not shooting by $B = 0$.
- *Endogenous variable D*: *D* represents the state of the dictator, $D = 1$ represents being fatally hit by a bullet, $D = 0$ represents not being hit by a bullet (i.e. survival).

These variables fit into the following causal graph:

The arrow from A to B models the influence that whether A shoots has causal influence on whether B shoots.

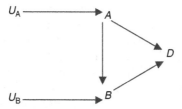

*Figure 2.7* Early pre-emption scenario

The causal information contained in Figure 2.7 may be expressed as follows: given that $U_1$ takes the value 1 and, if it were the case that $A = 1$ or $B = 1$, then it would be the case that $D = 1$. And, if it were the case that $A = 0$ and $U_B = 1$, then it would be the case that $B = 1$. Otherwise – that is, for all other combinations of the values of the variables – it is the case that $D = 0$.

What is the challenge of early pre-emption to an interventionist counterfactual theory? The fact that A shoots is the actual cause of the dictator's death. So, according to the interventionist theory of causation, the counterfactual conditional 'if there were an intervention such that A did not shoot, then the dictator would not have died' has to be true. Yet, in the example it is obviously not true, because assassin B would have killed the dictator if A – instead of B – had desisted from firing his gun. Fortunately, this example is only, at first sight, a counterexample to interventionism, because all conditions of our interventionist definition are satisfied and, therefore, A's shot $(A = 1)$ is an actual cause of the dictator's death $(D = 1)$. Let us examine the conjunction of three conditions of the definition of actual causation individually.

Pre-emption cases are usually understood to be scenarios of actual causation. Recall that Woodward's definition of (direct) actual causation states that $X = x$ is an actual cause of $Y = y$ iff the conjunction of the following conditions is satisfied:

*Actuality condition*: $X = x$ refers to the actually occurring event $c$ and $Y = y$ refers to the actually occurring event $e$.

*Redundancy range condition*: The values of all variables that are not on the directed path P from $X$ to $Y$ are in the redundancy range with respect to the actual values of variables on path P.

*Interventionist counterfactual dependence condition*: The interventionist counterfactual 'if there were an intervention such that the value of $X$ were changed from its actual value $x$ to some counterfactual value $x^*$, then the actual value $y$ of $Y$ would change to a counterfactual value $y^{*}$' is true.

So, let us check whether Woodward's definition of actual causation does the job of describing the scenario of early pre-emption.[22]

The actuality condition is satisfied, because A's shot and the dictator's death are actual events. The redundancy range condition and the interventionist counterfactual dependence condition are also satisfied: whether $A = 1$ causes the death of the dictator $(D = 1)$ depends on holding fixed or 'freezing'[23] other variables at their (actual) values that are not located on a causal chain (or a directed path) between the shot and the death. 'Holding fixed' a value of a variable means to set a variable to a certain (maybe actual) value with an intervention. The behaviour of assassin B is not on such a directed path from A's shot to the death of the dictator. So, in order to refute the counter-example, the interventionist has to rephrase the counterfactual in question with additional information in the antecedent: 'if there were an intervention such that A would not shoot and B would not fire his gun (as in the actual situation), then the dictator would not die' seems to be true in our example – or 'in our causal model', as we might say. In other words, the interventionist counterfactual dependence condition is satisfied. The redundancy range condition is satisfied, because – given that $A = 1$ – holding the value of $B$ fixed at its actual value ($B = 0$) does not change the actual value of $D$; namely $D = 1$. Therefore, the interventionist definition of actual causation is able to describe situations of early pre-emption correctly and this kind of situation is, consequently, no counter-example to interventionism.

*Late pre-emption*: Suppose that our two assassins[24] A and B have become enthusiastic about the successful elimination of the dictator of their own country. So, they become professional killers specialized in assassinating dictators all over the world. Today, they are invited to another country in order to kill an even crueller dictator there. This time, the actual events differ from the first assassination: A shoots first (at time $t_1$) and hits the dictator who dies instantly (at $t_2$). But, on this job, B *does* shoot (at $t_3$) one second after A's shot; B does *not* desist from firing his gun. B's bullet is a few seconds too late, because the dictator has dropped off the stage having already been hit by A's bullet. B's bullet just hits the wooden floor of the stage where the dictator had been standing when he was hit by A's bullet. Clearly, B would have killed the dictator if he had shot first instead of A.

We represent the situation by the same variables and their values as in the case of early pre-emption. Also, the causal graph (Figure 2.8) remains almost the same – except for the fact that we now explicitly consider temporal information about when an event takes place, and whether or not A shoots has no causal influence on B's decision to shoot: we can therefore omit the arrow from A to B.

The causal information contained in Figure 2.8, again, may be expressed as follows: given that $U_i$ takes the value 1 and, if it were the

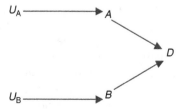

*Figure 2.8* Late pre-emption scenario

case that either $A = 1$ or $B = 1$, then it would be the case that $D = 1$. Otherwise – that is, for all other combinations of the values of the exogenous variables – it is the case that $D = 0$. In the scenario of late pre-emption, A's shot does not have any influence on B's shot.

What is the challenge to counterfactual theories of causation in the case of late pre-emption? In principle, the strategy required to describe the situation properly is the same as in the case of early pre-emption: A's shot is the actual cause of the dictator's death, but the counterfactual 'if there had been an intervention such that A had not taken the shot, then the dictator would not have died' is false, because in that case B would have shot the dictator instead (i.e. B's bullet would have hit and killed the dictator). The only difference between scenarios of early and late pre-emption consists in the fact that the potential cause of the death (B's shot) is pre-empted because the death has already occurred and not because – as in the early pre-emption scenario – the potential cause was cut off (i.e. B did not even fire) before the actual cause of the death occurred.

Again, we have to inquire whether we can describe the situation by means of our definition of actual causation. A's shot and the death of the dictator are actual events. Therefore, the first condition of actual causation holds. It can be easily shown that the interventionist coun-terfactual dependence condition is satisfied as well. We have to refor-mulate the counterfactual once more by holding fixed the values of the variable that represents the potential cause of the dictator's death: the counterfactual conditional 'if there had been an intervention such that A had not shot (at $t_1$) and B had not shot (at $t_1$) (as is actually the case), then the dictator would not have died (at $t_2$)' is true in the light of our scenario. If one holds $A$ fixed at its actual value 1 – that is, assassin A shoots, then changing the value of $B$ from its actual value 1 to a coun-terfactual value 0 does not change the actual value of $D$, namely $D = 1$. Therefore, the redundancy range condition is satisfied.

We may add that B's shot is a *potential, though pre-empted, cause* of the dictator's death, because the counterfactual 'if B were not to shoot (at $t_3$) and A were not to shoot (at $t_1$), then the dictator would not die at $t_2$ – instead, the dictator would die at $t_4$' is also true according to our example. However, B's shot at $t_3$ is not the actual cause of the dictator's death at $t_4$, because the dictator does not actually die at $t_4$ (he actually dies at $t_2$). Thus, the actuality condition is not satisfied for the statement 'B's shot at $t_3$ actually caused the dictator's death at $t_4$'.

In sum, scenarios of late pre-emption are not a threat to intervention-ism. Instead, we can explain why something is an actual cause even if a pre-empted, potential cause exists.

*Trumping pre-emption*: The case of trumping pre-emption is presented by Jonathan Schaffer (2004a: 67–70) as a counter-example to counterfactual theories of causation. The scenario of trumping pre-emption is supposed to show that there is a possible case in which there is causation without counterfactual dependence. One illustration of such a scenario is a chain of command in the army. Suppose that Corporal C receives orders from his superiors: Sergeant S and Sergeant S's superior Major M. Now, suppose that the major and sergeant both give the order to advance simultaneously. If so, the cause for the corporal to advance is the major's order but not the sergeant's order. The challenge of trumping pre-emption consists in the fact that the major's order is the actual cause of the corporal's advancing, but the counterfactual 'if the major were not to give the order to advance, the corporal would not advance' is false because the corporal would have advanced anyway, because of the sergeant's (actual) order to advance.

Woodward (2003: 81f.) suggests a way to deal with trumping. In order to describe the trumping case in the framework of the interventionist theory, it is crucial to determine correctly the range of the variables involved and to re-describe the situation. Let variable $M$ represent the major's order by ranging over a set of three possible values {order to advance; order to retreat; no order at all}. Similarly, let $S$ represent the sergeant's order, ranging over a set of three possible values {order to advance; order to retreat; no order at all}. That is, $S$'s range is identical with $M$'s range. $C$ is a variable representing the action of the corporal who takes the major's orders and the sergeant's orders, ranging over the set of possible values {advance; retreat; do nothing}. How does this choice of variables help to describe the scenario of trumping pre-emption correctly? An interventionist has to show that the major's order to advance actually caused the corporal's advance. So, the question is whether the interventionist definition of actual causation leads to the right result. In agreement with the actuality condition, the statements $M$ = *order to advance* and $C$ = *advance* are true in the actual world. According to the interventionist counterfactual dependence condition, one can hold fixed the value of $S$ at some non-actual value. This can be expressed by interventionist counterfactuals such as 'if there had been an intervention such that $M$ = *no order at all* and $S$ is held fixed at *do nothing*, then it would have been the case that $C$ = *do nothing*' and 'if there had been an intervention such that $M$ = *order to retreat* and $S$ is held fixed at the actual value *order to advance*, then it would have been the case that $C$ = *retreat*'. Woodward takes these counterfactuals to be true, and this judgement seems to be in perfect agreement with Schaffer's original description of the case. If this is the case, then – contra Schaffer – the corporal's action does counterfactually depend on the major's orders. According to both the interventionist counterfactuals given, the values of $S$ are in the redundancy range with respect to $C$ – that is, changing the value of $S$, given that the actual value is assigned to $M$ – has no influence on the value of $C$. Thus, the redundancy range condition is satisfied. In sum, Woodward's definition refutes the scenario of trumping pre-emption as a counter-example.

*Symmetric over-determination*: Suppose that our assassins A and B shoot and hit the target at the same time (cf. Woodward 2003: 82f.). What is the intuitive judgement in this case? The intuition is that both shots are individually and equally to be counted as causes of the death of the dictator. That is, $A = 1$ (i.e. A shoots) is an actual cause of $D = 1$ (the dictator's death) and $B = 1$ (B shoots) is an actual cause of $D = 1$. Analogously to other pre-emption cases, counterfactual dependence does not seem to be necessary for causation, because the counterfactuals 'if $A = 0$ were the case, then $D = 0$ would be the case' and, respectively, 'if $B = 0$ were the case, then $D = 0$ would be the case' are false, because there is an over-determining cause to the effect that the dictator dies (i.e. $D = 0$). So, are interventionists able to describe symmetric over-determination correctly? Let us first examine whether $A = 1$ is an actual cause of $D = 0$. According to the actuality condition, $A = 1$ and $D = 0$ are actually true statements. Holding $B$ fixed at its counterfactual value 0 yields a counterfactual dependence of $D$ on $A$. This dependence can be stated in the interventionist counterfactual 'if it were the case that there is an intervention such that $A = 0$ and $B = 0$, then it would be the case that $D = 0$'. To think that this counterfactual is true is to agree with the scenario of symmetric over-determination. Thus, the interventionist counterfactual dependence condition is satisfied. Holding $B$ fixed at its value 0 meets the redundancy range condition, because the fact that $B = 0$ does not alter the value of $D$ given that actual value of $A$. Analogously, one can show that $B = 1$ is an actual cause of $D = 1$ by holding $A$ fixed at its value 0. Therefore, one can conclude that symmetric over-determination is not a counter-example to interventionist theories of causation.

In general, we can conclude that Woodward's theory of causation can cope with various scenarios of pre-emption. Thus, the interventionist theory satisfies the distinction criterion of adequacy with respect to this class of causal scenario. Now, let us turn to another kind of counter-example concerning the transitivity of causation.

*Transitivity of causal relations*

In the current debate on causation, philosophers seem to converge on the opinion that it is problematic to assume that causation is transitive. Let us take a step back and wonder: why should one worry about scenarios revealing the non-transitivity of causal relations? Why should one consider these scenarios to be a dangerous counter-example at all? Recall the beginning of this section: some philosophers that defend a counterfactual theory of causation – most prominently, David Lewis and his followers – build explicitly into the definition of actual causation that causation is a transitive relation. Note that the main motivation for David Lewis (1973b, 2004) to build transitivity directly into the definition of causation is to deal with pre-emption scenarios. So, the assumption of transitivity is part of a strategy of defending a Lewisian counterfactual theory that strongly differs from interventionism. Interventionists are not committed to assuming transitivity for two reasons: transitivity is not directly assumed in our definition of actual causation; and interventionists are not indirectly committed to transitivity by defending their approach against counter-examples. Contrary to Lewis, interventionists do not need to assume transitivity to model situations of early, late, and trumping pre-emption, and symmetric over-determination. Therefore, they have no further independent reason to tie causation tightly to transitivity.[25] It seems that scenarios indicating the intransitivity of causation are merely a counter-example to a defensive strategy proposed by Lewis to defend his own theory. I consider it to be a strength and a virtue of the interventionist definition of causation that it is not committed to the claim that causation is transitive (cf. Hitchcock 2001).

These situations, which appear to be counter-examples to Lewis's theory, can be adequately dealt with in an interventionist framework (cf. Hitchcock 2001; Woodward 2003: 79–81). Causation is transitive if it is true for three events $c$, $d$ and $e$ that, if $c$ causes $d$ and $d$ causes $e$, then $c$ causes $e$. In recent discussions,[26] thought-experiments have been invented that render the thought implausible that causal dependencies are transitive. A possible worry could be that it is important to stress that making up peculiar cases in which causal dependence is not transitive certainly does not show that transitive causal relations are unusual or rare: quite the contrary. It seems that most causal relations are transitive; that is, normally causation is transitive. The point is, rather, that we should not build the feature of transitivity directly into the definition of actual causation, because – according to this type of counter-example – causation is not essentially a transitive relation. However, this worry is ill-founded: the meaning of 'transitivity' which is at issue in the debate *is* the essential reading. That is, one single example of a situation where causation is not transitive is sufficient to claim that causation is not transitive.

Imagine the following situation (Hall 2004): a boulder is dislodged somewhere in the mountains (event $c$), close to hiker A's whereabouts.

Hiker A sees the boulder, ducks early enough and is able to hide in a cave (event *d*). Hiding in this cave allows her to survive (event *e*); otherwise she would have certainly been killed. If one assumes, in agreement with Lewis, that causation is transitive, then one has to conclude that the boulder being dislodged causes A's survival. However, it is certainly hard to admit that this is a proper description of the causal history in the example.

We represent the situation by choosing the following variables and their values:

- *Exogenous variables $U_i$*: A variable $U_1$ might represent the mental states of hiker A (e.g. that she has the desire to hike and to survive her hiking trip, her belief that it is a perfect day to go hiking and so on). $U_1 = 1$ expresses that hiker A is in those mental states; $U_1 = 0$ means that she is not. We may also consider a variable $U_2$ that represents some relevant geological and climatic factors that cause the boulder to roll downhill. $U_2 = 1$ means that these factors do cause the boulder to roll; $U_2 = 0$ represents the absence of these factors. The $U_i$ are set to value 1.
- *Endogenous variable B*: Variable *B* represents the behaviour of the boulder. $B = 1$ represents that the boulder is dislodged; $B = 0$ means that it is not.
- *Endogenous variable D*: Variable *D* represents the event of the hiker's ducking. $D = 1$ represents the fact the hiker ducks; $D = 0$ represents that she does not.
- *Endogenous variable S*: Variable *S* stands for the survival of the hiker. $S = 1$ means that she survives; $S = 0$ represents that she does not.

The causal graph in Figure 2.9 represents the variables and their values and the causal structure of the scenario.

The causal information contained in the causal model underlying the graph in Figure 2.9 may be expressed as follows: given that $U_i$ takes the value 1,

- if it were the case that $B = 1$ and $D = 1$, then it would be the case that $S = 1$

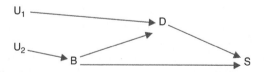

*Figure 2.9*  Boulder scenario

- if it were the case that $B = 0$ or $D = 1$, then it would be the case that $S = 1$
- if it were the case that $B = 0$ and $D = 0$, then it would be the case that $S = 1$
- if it were the case that $B = 1$ and $D = 0$, then it would be the case that $S = 0$.

The crucial question is whether we are forced to admit that, according to the interventionist definition of actual causation, the fact that the boulder is rolling downhill towards the hiker (i.e. $B = 1$) is the actual cause of the survival of our hiker (i.e. $S = 1$).

Although some philosophers attempt to defend transitivity (cf. Lewis 2004), one might simply accept the counter-example against the claim that causation is transitive. It seems natural for interventionists to do so, because they do not have to care about transitivity, and it is not required in the interventionist definition of actual causation that causation has to be transitive. Interventionists have to show that their definition of actual causation does not apply to the boulder scenario. In accordance with our intuitions, the boulder going down the mountain is not the actual cause of the hiker's survival, because our definition of actual causation is not satisfied for these two events. Let us examine whether this is true.

The actuality condition is fulfilled, because the boulder going down the mountain and the survival of our hiker are actually occurring events. What about the interventionist counterfactual dependence condition? If there were an intervention such that the value of $B$ were to change from $B = 1$ to the counterfactual value $B = 0$, and given that we hold $D$ fixed at its actual value $D = 1$, then the value of $S$ would not change. If that is correct, then there is no counterfactual dependence of the survival on the falling of the boulder, because the following two counterfactuals are true: first, if there were no boulder rolling down the mountain and our hiker were to duck, then it would be the case that the hiker survives. Second, if there were a boulder on its way down the hill and our hiker were to duck, then it would be the case that the hiker survives. Therefore, the interventionist counterfactual dependence condition is not satisfied. It follows that $B = 1$ cannot be an actual cause of $S = 1$. One might wonder whether the redundancy range condition might have been violated. But this is not the case. The crucial interventionist counterfactuals hold $D$ fixed at its actual value 1. The fact that $D = 1$ is clearly in the redundancy range, because the fact that $D = 1$ does not alter the actual value of $S$ (i.e. $S = 1$) conditional on the fact that $B$ takes its actual value (i.e. $B = 1$). Thus, the interventionist is able to state clearly that the boulder is not an actual cause of A's survival.

Consequently, the interventionist definition of actual causation is not satisfied in the example case. In other words, the boulder being dislodged is not the actual cause of the survival of the hiker. Therefore, interventionists can state a straightforward denial that causation is transitive.

### Results: a success for interventionists

Let us take stock. What is the general lesson to be learned from the alleged counter-examples that have been presented and discussed? One can be confident that interventionists make convincing attempts to deal with various pre-emption scenarios. Interventionists also argue convincingly that they are not committed to the problematic claim that causation is transitive. This is a great success for interventionist theories of causation, because they can account for causal scenarios that remain counter-examples to the Lewisian brand of the counterfactual theory of causation. Based on these results, it seems fair to conclude that interventionist theories of causation meet the distinction criterion of adequacy to a satisfying degree. Obviously, it has not been possible to discuss all the counter-examples in the debate. Convincing attempts to reject various further counter-examples can be found in Hitchcock (2001), and Halpern and Pearl (2005).

## Conclusion

This chapter introduced Woodward's interventionist theory of causation. The reconstruction of his view comprised the definition of the notions of an intervention variable and of an intervention, as well as the definitions of various causal notions (such as direct and indirect type-level causation, actual causation and so on). The formal background of interventionist theories was presented, most importantly causal graphs and Bayesian networks. The primary use of this formal background was in order to distinguish methodological and semantic questions concerning causation. It was suggested that a discussion of Woodward's theory of causation does not have to address the methodological questions and the problems related to answering these methodological questions. Subsequently, I discussed whether the interventionist theory is an adequate theory of causation for the special sciences. One can preliminarily conclude that interventionist theories of causation at least promise to meet the naturalist criterion and the distinction criterion of adequacy. By virtue of this fact, interventionist theories seem to be a good candidate for an adequate explication of causation in the special sciences.

# Part II
# What is Wrong with Interventionist Theories

# 3
# Counterfactuals: A Problem for Interventionists?

## Interventionist theories of causation involve counterfactuals

Part I set the stage for a discussion of a theory of causation in the special sciences. Chapter 1 argued for the naturalist criterion and the distinction criterion of adequacy for an explication of causation in the special sciences. Chapter 2 presented Woodward's interventionist theory of causation and its virtues were highlighted. Part II will focus on objections to the interventionist theory of causation. This chapter will raise a problem for interventionists that stems from the meaning of counterfactual conditionals.

Interventionists claim to advocate a (modified) counterfactual theory of causation, because their definitions of causation depend essentially on counterfactual conditionals (cf. Woodward 2003, also Hitchcock 2001, Halpern and Pearl 2005). More precisely, interventionists appeal to counterfactuals of the following form:

> If the value of $X$ were changed to be $x_i$ by an intervention $I = i$, then the value of $Y$ would change to $y_k$. (cf. Woodward 2003: 15)

Call counterfactuals of this form 'interventionist counterfactuals'. Counterfactuals of this form are built into interventionist theories of causation. For instance, recall Woodward's definition of a direct type-level cause:

> A necessary and sufficient condition for $X$ to be a direct cause of $Y$ with respect to some variable set **V** is that there be a possible intervention on $X$ that will change $Y$ (or the probability distribution of $Y$)

when all other variables are held fixed at some value by intervention. (Woodward 2003: 55)

So, according to Woodward, if $X$ is a direct cause of $Y$, then there is a *possible intervention* on $X$ that changes $Y$. What is the *counterfactual* underlying this definition of a direct cause? It is an interventionist counterfactual, which is formulated in terms of random variables[1] $X$ and $Y$ (being elements of **V**) and their range of possible values $\{x_1, ..., x_n\}$ and $\{y_1, ..., y_k\}$,[2] which Woodward also calls interventionist or 'active' counterfactuals (cf. Woodward 2000: 199; 2003: 122):

If the value of $X$ were changed to be $x_i$ by an intervention $I = i$ and all other variables (in a given causal model) are held fixed at some value by intervention, then the value of $Y$ would change to $y_i$.

The antecedent proposition $X = x$ is conceived as the outcome of an intervention on the variable(s) referred to in the antecedent. Woodward's notion of an intervention has been reconstructed in detail (Chapter 2, pp. 29–32). The main results were that Woodward defines interventions via the notion of an intervention variable. According to Woodward (2003: 98), a variable $I$ is an intervention variable for $X$ (relative to $Y$) iff:

- $I$ is a cause of $X$;
- there is at least some value of $I$ such that if $I$ takes this value, then $X$ depends only on $I$ and $X$ depends on no other variables; that is, $I$ is the only cause of $X$;
- $I$ is not a direct cause of $Y$, and if $I$ is a cause of $Y$ then $I$ is an indirect cause of $Y$ via a causal path leading through $X$ and a – possibly empty – set of intermediate variables $Z_1, ..., Z_n$;
- $I$ is probabilistically independent of other causes $W_1, ..., W_n$ of $Y$, which are not on a causal path leading from $X$ to $Y$;
- $I$ does not alter the relationship between $Y$ and its causes $W_1, ..., W_n$, which are not on a causal path leading from $X$ to $Y$.

The notion of an intervention variable is used to define the notion of an intervention:

Any value $i_i$ of an intervention variable $I$ (for $X$ relative to $Y$) is an intervention on X relative to Y iff it is the case that the value of X counterfactually depends on the fact that $I$ has the value $i_i$. (cf. Woodward 2003: 98)

As shown (Chapter 2, p. 32), the notion of a possible intervention and interventionist counterfactuals are the building blocks for an interventionist theory of causation.

Interventionists claim to advocate a counterfactual theory of causation. However, although interventionists assert that the meaning of counterfactuals is fixed by truth conditions, they fail to provide truth conditions for counterfactual conditionals. The lack of an account of the meaning of counterfactuals is a desideratum of the interventionist theory of causation. This becomes obvious if one contrasts the interventionist approach with a competing counterfactual theory of causation such as Lewis's theory. In order to solve this problem, one has to provide a semantics for counterfactuals. Since Lewisian semantics has become orthodoxy in the debate, it would be easiest to adopt this standard account of semantics. However, as will be reconstructed, interventionists have presented convincing arguments against Lewisian semantics. Taking these arguments seriously, this chapter will offer three alternative standard theories of the meaning of counterfactuals that might suit the interventionists' demands: one can modify Lewisian possible worlds semantics; one can use a meta-linguistic, or Goodmanian, account; or one can adopt a suppositional theory of pragmatic meaning.

One important note should be added about the dialectic role of this chapter. Proposals for various options for interventionist semantics will accept the notion of a possible intervention only preliminarily and only for the sake of the argument. Chapters 4, 5, and 6 will argue that the notion of a possible intervention and the interventionist semantics are deeply problematic.

## A problem for interventionism: truth conditions for counterfactuals

Although interventionism is supposed to be a counterfactual theory of causation, surprisingly its proponents do not provide an account of truth conditions for counterfactuals. Note that this problem also infects theories of laws that rely on interventionist counterfactuals (cf. Woodward and Hitchcock 2003; Leuridan 2010), and mechanist theories of explanation that essentially rely on an interventionist explication of causation (cf. Glennan 2002, Craver 2007). Craver (2007: 224–7) is one of the few people in the debate that acknowledges the problem (although without offering a solution).[3] This is a chronic problem, because any proponent of a counterfactual theory seems to be required to have a semantics for counterfactuals. However, instead of providing truth conditions,

Woodward sometimes argues for an entirely different issue; namely, that one might gain knowledge of the truth of counterfactuals by carrying out (hypothetical) experiments (cf. Woodward 2003: 70–4).[4] This might well be true, but it is not the same as giving truth conditions. Moreover, there appears to be a tension between using counterfactuals to explicate causation and not accounting for the meaning of counterfactuals, because Woodward describes his own project as clarifying the meaning or content of causal notions (cf. Woodward 2003: 38, see also Chapter 2, pp. 32–5). Woodward seems to presuppose that the meaning of a sentence consists in its truth conditions (cf. Woodward 2003: 7–9, 2008: 193–6; Strevens 2008: 184). The task of providing these truth conditions becomes even more pressing, because Woodward argues for objective truth-values of counterfactuals. In his own words:

> We think instead of [a counterfactual] as having a *determinate meaning and truth value* whether or not the experiment is actually carried out—it is precisely because the experimenters want to *discover* whether [this counterfactual] is true or false that they conduct the experiment. (Woodward 2003: 123, emphasis added)

Woodward claims that his theory of causation requires that:

> there be facts of the matter, independent of facts about human abilities and psychology, about which counterfactual claims about the outcome of hypothetical experiments are true or false and about whether a correlation between $C$ and $E$ reflects a causal relationship between $C$ and $E$ or not. (Woodward 2003: 122f.)

In other words, Woodward tells us that there are truth conditions and mind-independent truth-makers for counterfactuals. However, he does not tell us what precisely these truth conditions are. This is a rather odd desideratum for interventionists, because Woodward and other interventionists claim that:

- causal statements and counterfactual claims have meaning;
- the meaning of causal and counterfactual statements is determined semantically by truth conditions;
- stating truth conditions, one needs to invoke causal notions.

It is striking that Woodward (together with Christopher Hitchcock) seems to be aware of the fact that there is a problem. Although

Woodward does not succeed in solving the problem entirely, he points out two strategies with which to deal with stating the truth conditions of interventionist counterfactuals.

First, he claims that his account of causation needs to be supported by a theory of truth conditions for counterfactuals. Responding to Quine's objection that counterfactuals 'lack a clear meaning or that their truth conditions are so vague and context dependent that they not suitable for understanding [...] any notion that might be of scientific interest' (Woodward 2003: 122), Woodward defends the following view:

> A manipulability [i.e. interventionist] framework for understanding causation provides a response to this worry [i.e. the worry that counterfactuals lack truth conditions]. It suggests that the appropriate counterfactuals for elucidating counterfactual claims are not just any counterfactuals, but rather *counterfactuals of a very special sort: those that have to do with the outcomes of hypothetical interventions.* (Woodward 2003: 122, emphasis added)

All that Woodward does here is to restrict the attention to a certain class of counterfactuals: interventionist counterfactuals. Unfortunately, this mere restriction by Woodward is not sufficient for providing truth conditions for counterfactuals. However, his semantic proposal seems to be that counterfactuals are true if the antecedent is understood to be the outcome of an intervention and the consequent is true. This basic idea of an interventionist semantics will be spelled out in this chapter (pp. 78–9).

Second, Woodward also indirectly commits himself to the task of providing truth conditions for counterfactuals because he criticizes Lewis's semantics for counterfactuals extensively (cf. Woodward 2003: 138–45). This critique and the claim that there are truth conditions of (at least, interventionist) counterfactuals create a dialectic obligation to provide a semantics for counterfactuals that is an alternative to Lewis's semantics. Unfortunately, Woodward's self-acknowledged commitment to truth conditions is not followed by an explicit and satisfactorily detailed account of truth conditions. Furthermore, Woodward is often ambiguous with respect to the need to state the truth conditions of counterfactuals. On the one hand, Woodward sometimes appears to be prepared to accept the truth conditions as primitive and to engage only in 'pragmatic considerations' (Woodward 2004: 45f.) of the use of counterfactuals; that is, (a) their use in conceptual analysis of other concepts (e.g., the concepts of law and causation), (b) how they can be tested empirically, (c) how (i.e. according to which rules) we can use them in inferences. On the other hand, it seems that Woodward is eager to develop a proper interventionist semantics for counterfactuals. Concerning the second strategy, one can find several attempts to take up the obligation in Woodward's papers. Unfortunately, Woodward's attempts to state truth conditions of counterfactuals (essentially, by relying on the notion of an intervention) are not systematic: rather, they have the nature of asides. However, a couple of essential features of interventionist semantics can be identified quite clearly in two quotes:

> A counterfactual of the form 'if $X$ were to have the value $x$, then $Y$ would have the value $y$' *is true if and only if $Y$* has the value $y$ in the hypothetical situation (or possible world) where (i) the value of $X$ is equal to $x$; and (ii) all other variables have their actual values with the exception of [the intervention variable] $I$, and any variable for which $I$ is causally relevant, where $I$ is an intervention variable for $X$ with respect to $Y$. (Woodward and Hitchcock 2003: 13f, emphasis added)

Elsewhere, Woodward characterizes the meaning of a counterfactual with respect to laws (i.e. invariant generalizations):

> If counterfactual claims are to be used in a disciplined way, [...] this is accomplished by the characterization of the notion of an intervention and the use of systems of equations or functional relationships, or by means of directed graphs, together with precise rules for transforming these to describe non-actual hypothetical situations of various sorts. The equations and graphs, together with the rules specifying what happens under interventions, tell us explicitly just what should be regarded as changed and what should be 'held fixed' when we entertain various counterfactual assumptions. *If the equations and graphs are correct, they enable us to determine what would happen under these assumptions.* (Woodward 2004: 44f, emphasis added; also cf. Woodward 2003: 281; Hitchcock 2001: 280–3)

According to the first quote, the counterfactual 'if $X$ were to have the value $x$, then $Y$ would have the value $y$' is semantically evaluated by referring to other possible worlds where an intervention on $X$ exists. According to the second quote, the consequent can be inferred from 'equations' and the antecedent, if the counterfactual is true. These two quotes are problematic in the following respect: they are too ambiguous to count as a complete interventionist semantics. Concerning the first quote, it not immediately clear why worlds in which interventions on the antecedent variable exist should be used to evaluate counterfactuals. Concerning the second quote, one needs further explanation of how exactly 'equations' enable us to evaluate counterfactuals. As I will argue, the first and the second quote pull in different directions: the first suggests a Lewisian possible worlds semantics. The second suggests either a Goodmanian meta-linguistic semantics or a suppositional view of the meaning of counterfactuals.

The result is that, on the one hand, interventionists claim to provide a version of a counterfactual theory of causation. But, on the other hand, they lack a convincing account of truth conditions for counterfactuals. Obviously, this outcome is an uncomfortable situation for proponents of a counterfactual theory of causation. There are at least two options to deal with the situation: either one denies that there is a problem at all, or one accepts the fact that the situation is challenging. One way to follow the strategy of denying there is any problem is to agree with Woodward's claim that all we need to know is that there are truth conditions for counterfactuals. But which truth conditions these are is of little interest. This view could be understood as some kind of primitivism regarding the truth conditions of counterfactuals. What is interesting is the methodological question; that is, how one confirms counterfactual statements. A natural solution to the perception that the situation is challenging would be to give another, more precise account of the meaning of counterfactuals than the ambiguous attempts to provide an interventionist semantics already introduced and briefly discussed.

This chapter explores the second option, and will consider three standard approaches to the meaning of counterfactuals: possible worlds semantics, the meta-linguitic account, and the suppositional view.[5] The overall goal of this chapter is constructive: it aims to strengthen the interventionist theory of causation by supplementing it with an account of meaning for interventionist counterfactuals. However, a semantics will be provided on behalf of the interventionists, purely for the sake of argument.

## The easy way out: standard possible worlds semantics

The orthodox solution to the interventionist problem of finding a semantics is to adopt *the* standard semantics for counterfactuals. In many of the current philosophical debates, the standard semantics is David Lewis's possible worlds semantics.

Before introducing Lewis's semantics, I want to restrict the semantics to formulas of a certain kind. A semantics for those counterfactuals, which are supposed to express causal information in the interventionist framework, has only to deal with sentences of the object language that express the proposition that some event-type occurs. The interventionist analogue to this kind of proposition consists in propositions that a variable takes one of its possible values, such as $X = x$ (see Chapter 2, pp. 26–7, for details).[6] Additionally, the antecedent proposition expresses that a variable is set to a value by an intervention. In his theory of causation, Lewis originally considered only the sentence expressing the qualitative proposition that an event $e$ occurs. In one of his last papers on causation, Lewis can be taken to have converted to variable expression when he introduces the sentences that express propositions about 'not too distant alterations' of an event (cf. Lewis 2004: 91f.). In other words, Lewis and Woodward converge on the syntax of causal formulas.[7] In what follows, Greek letters will be used as meta-variables for sentences that express the proposition that a variable takes a certain possible value.

According to Lewis's possible worlds semantics, a counterfactual conditional is true in a world w iff the following conditions hold:

> *Lewisian semantics*: The counterfactual *if φ were the case, then ψ would be the case*[8] is (non-vacuously) true at w if and only if ψ is true in the closest φ-worlds. (Lewis 1973a: 164; 1973b: 16)

According to Lewis, the closeness relation is characterized by two important features: first, the closeness relation is a weak ordering relation. That is, (a) a world w might be equally close to two worlds u and v, and (b) any two worlds are comparable with respect to their closeness. Second, the evaluation world w is the closest world to itself. Closeness is measured by criteria of similarity. Similarity consists mostly in shared laws of nature and the amount of shared facts (or, property instantiations in space–time regions, as Lewis puts it) in each world. According to Lewis, we are supposed to employ the following weighted criteria as a similarity heuristic[9] to select the closest possible worlds:

(1) It is of the first importance to avoid big widespread, diverse violations of law.
(2) It is of the second importance to maximize the spatio-temporal region throughout which a perfect match of particular fact prevails.

(3) It is of the third importance to avoid even small, localized, simple violations of laws.

(4) It is of little or no importance to secure approximate similarity of particular fact, even in matters that concern us greatly (Lewis 1979: 47f.).

Lewis chooses the term 'miracle' to refer to violations of actual laws in other possible worlds. 'Big miracles' refer to possible worlds in which the actual laws are violated to such an extent that one is inclined to say that big-miracle worlds simply have different laws than the actual world. 'Small miracle' denotes events in possible worlds in which the actual laws have a local exception at a time shortly prior to the antecedent-event. An instructive way to understand a small miracle is to regard it as setting new, local initial conditions. Given that these new initial conditions obtain, a small-miracle world conforms to the same laws as before the miracle. Lewis's idea is that, if the world evolves in accord to the pre-miracle laws, then the antecedents event occurs shortly after the miracle. After the miracle, a determinstic world takes a different course than the actual course of events. 'Perfect match' refers to similarity in particular facts. The upshot is that laws and facts are the essentials ingredients of Lewisian semantics.

To complete the picture of possible worlds semantics in its standard form, Lewis considers possible worlds to be entities that are just like the actual world. What matters for our concerns is not mainly the metaphysics of possible worlds (i.e. the question as to whether worlds are concrete entities as Lewis supposes, or linguistic ones and so on). What does matter, rather, is that, according to Lewis, worlds (and especially the closest worlds) are as detailed as our world (i.e. the real spatio-temporal entity that is the world we live in, cf. Hüttemann 2004: 113). Lewis illustrates this idea by evaluating the counterfactual 'If kangaroos had no tails, they would topple over' for the actual world. Asking the question 'What are the closest worlds?', we learn a good deal about what Lewisian worlds are like:

We might think it is best to confine our attention to worlds where kangaroos have no tails and everything else is as it actually is; but there are no such worlds. Are we to suppose that kangaroos have no tails but that their tracks in the sand are as they actually are? Then we shall have to suppose that these tracks are produced in a way quite different from the actual way. [...] And so it goes; respects of similarity and difference trade off. (Lewis 1973a: 9)

Let us call such detailed worlds 'big worlds'. It is important to notice Lewis's affinity to big worlds, when it comes to the critique of Lewisian semantics in the following sections.

To sum up, according to Lewisian semantics, one cannot provide truth conditions for counterfactuals without relying on laws. Contrary to the ambiguity in interventionist semantics, laws and particular facts (although not interventions) are the decisive building blocks of the semantics.

## The interventionist critique of possible worlds semantics

Some people are quite unorthodox: interventionists refuse to adopt Lewisian semantics. Let us reconstruct three objections against the standard semantics raised by interventionists: the objection against Similarity I, the objection against Similarity II, and the objection against 'big worlds'.

### Objection against Similarity I

Interventionists object to Lewisian semantics in that the similarity relation is arbitrary and imprecise. The choice and order of the similarity criteria seems arbitrary, because Lewis offers no justification. Further, even if one agreed on the choice and the preference order of similarity criteria, there is still a problem about whether these criteria enable us to determine the closest possible worlds. Woodward makes this point as follows:

> Does any gain in perfect match, no matter how minimal [...] count for more than any localized violation of a law? Moreover, how do we count miracles for the purpose of deciding whether we have a single small miracle or a big miracle? Lewis says that big miracles are made of many small ones, but how many is 'many'? (Woodward 2003: 138)

These objections amount to an argument against the similarity heuristics: in order to evaluate a counterfactual, one has to determine the truth conditions in a non-arbitrary and precise way. But they do not seem to be determinable in this way. Lewis might object to this observation: (a) whether a semantics is successful depends on whether it evaluates counterfactuals adequately as true or false in possible situations, and (b) the fact that the criteria of similarity are vague is intended to capture the context dependence of the truth conditions of counterfactuals. Lewis might argue that this condition of adequacy can be met by a semantics even if (and maybe just because of the fact that) the truth conditions are somewhat arbitrary and imprecise. Consequently, Woodward does not stop his critique by pointing out that the similarity heuristic is arbitrary and imprecise. He continues to deny that a possible worlds semantics plus a similarity heuristic yields adequate results. In support of this line of argument, he presents the following two counter-examples against Lewis's similarity heuristic.[10]

Woodward's first counter-example presupposes the following situation:

> You are driving on an unfamiliar freeway in the left-hand lane when, unexpectedly, the exit you need to take appears on the right. You are unable to get over in time to exit and as a result are late for your appointment. There are only two lanes, left and right. Driving in the left-hand lane (rather than the right) caused you to be late. (Woodward 2003: 142)

In this situation, the counterfactual 'if you had not been driving in the left lane (that is, if you had been driving in the right lane), then you would not have been late' is supposed be evaluated as true. Woodward argues that the counterfactual comes out as false if one follows Lewis's similarity heuristic. According to Lewis's similarity heuristic, the closest antecedent-worlds (where the driver is in the right lane as the antecedent of the counterfactual states) are worlds that (a) perfectly match the actual world until time *t* when the driver is close to the exit (and in the left lane), and where (b) a small miracle occurs at *t* such that the car is in the right lane. At time *t* a small miracle happens: 'your car dematerializes and reappears instantaneously in the right hand lane just before the exit' (Woodward 2003: 142). According to Lewis, the counterfactual is true iff the driver is not late in these closest antecedent-worlds (as characterized above). As Woodward observes, this is not the case, because the allegedly small miracle causes events that (are likely to) prevent the driver being on time for their appointment (i.e. not late as the consequent of the counterfactual states):

> [I]t is not at all unlikely that the very occurrence of this miracle will produce effects that will interfere with your exiting. For example, other drivers will be startled and distracted by the sudden appearance of a car in the right-hand lane, perhaps very close to or in contact with cars that already occupy the right lane. Perhaps a collision will

occur or other drivers may swerve or slow down, with the result that your exit is impeded. There will also be a great rush of air into the space in the left lane previously occupied by your car, a similar rush as air is displaced from the right-hand lane, and accompanying loud noises, all of which may also interfere with your exit. *So, if this is the relevant world to consider,* [the counterfactual] *may well be false, contrary to the result that we want.* (Woodward 2003: 142)

This is a counter-example to Lewis's similarity heuristic, because the heuristic leads to an evidently inadequate evaluation of a counterfactual (cf. Bennett 2003: ch. 13, for a discussion of structurally analogous examples against Lewis's similarity heuristic).

Woodward (2003: 139f.) also discusses a second counter-example to Lewis's similarity measure. This counter-example is supposed to show that Lewis's similarity heuristic picks out the wrong worlds in order to describe a specific type of causal structure. Suppose that, at $t_0$, $X$ directly and independently causes the occurrence of each of $E_1$, ..., $E_n$ at $t_1$ and of $Y$ at $t_2$. According to Woodward, the counterfactual 'if $E_1$, ..., $E_n$ had not occurred, then $Y$ would not have occurred' should be evaluated as false, because $E_1$, ..., $E_n$ are described as irrelevant for the occurrence of $Y$. Woodward claims that, according to the similarity heuristic, the closest possible worlds are such that: (a) $X$ occurs at $t_0$, and (b) $n$-1 miracles occur at a point in time between $t_0$ and $t_1$ in order to prevent $E_1$, ..., $E_n$ from happening. In these worlds, the counterfactual is false as it should be – given that $X$ causes $Y$, and given that $E_1$, ..., $E_n$ do not cause $Y$. Woodward observes a problem: if $n$ is large (i.e. if $X$ has a large number of effects at $t_1$), then the closest worlds contain many 'small' miracles expunging $E_1$, ..., $E_n$ that add up to a big miracle. But this conflicts with the most important criterion of Lewis's similarity heuristic. Thus, according to Lewis's similarity heuristic, worlds as characterized above cannot be the right ones in which to evaluate the counterfactual. Woodward objects that 'intuitively' worlds with $n$-1 miracles seems to be the 'correct worlds' in order to evaluate the counterfactuals. Let us call these worlds many-miracle worlds. Yet, there are worlds that should be considered to be closer according to the similarity heuristic: worlds where only one miracle occurs to expunge $X$ and, consequently, also its effects $E_1$, ..., $E_n$. The trouble is that this one-miracle world is a non-$Y$ world. In other words, the counterfactual 'if $E_1$, ..., $E_n$ had not occurred, then $Y$ would not have occurred' is true, although it should be evaluated as false. Thus, the similarity heuristic again leads to an inadequate result. Woodward concludes that it seems to be an ad hoc strategy to reject the natural appeal to many-miracle worlds in order to vindicate Lewis's similarity heuristic.

## Objection against Similarity II

Interventionists complain that the criteria for similarity are not only imprecise and arbitrary, but they are also subject to a further problem: nothing in scientific practice corresponds to Lewis's criteria of similarity. As Hitchcock and Woodward formulate the point: 'We doubt that there is anything in *scientific* practice that tells us how to count miracles or to make such comparisons in a non-arbitrary way' (Woodward and Hitchcock 2003: 9, emphasis added). However, interventionists want semantics to be a *naturalized* explication or a rational reconstruction of scientific practice. Therefore, they conclude that one cannot be satisfied with Lewisian semantics (cf. Woodward and Hitchcock 2003: 9).

Note that both objections against similarity focus on the similarity heuristics only. In order to stress this point, note that the preceding objections raised by interventionists challenge neither possible worlds semantics for counterfactuals (relying simply on a vague notion of closeness) nor the logic of counterfactuals. In principle, the heuristic of similarity as presented by Lewis might be rejected, while one may still (partly or entirely) maintain the possible worlds semantics and the logic of counterfactuals that Lewis has introduced (cf. Lewis 1973a: Chapter 6). This logic implies that there is a weak ordering relation ranging over possible worlds. This implication does not force one to assume that this ordering is best understood in terms of Lewis's similarity heuristic.

## Objection against 'big worlds'

In the framework of his possible worlds semantics, Lewis refers to big worlds. Again, interventionists want semantics to be an explication or a rational reconstruction of scientific practice. But, in the sciences, one finds only very restricted (i.e. abstract and/or idealized) mathematical models of phenomena.[11] Restricted models have an impact on how one pictures possible worlds to be: a possible world is[12] a specific assignment of values to the variables in a model. These model worlds are not Lewisian big worlds – they are small, restricted worlds. Following the rule that semantics should pursue an explication or a rational reconstruction of scientific practice, interventionists conclude that assuming big worlds should be avoided.

According to Kripke, the sciences merely stipulate small worlds consisting of the possible values that the endogenous and exogenous variables in a (very restricted) model take. A classic illustration of small worlds by Kripke may help to clarify this point:

> Two ordinary dice (call them A and B) are thrown, displaying two numbers face up. For each die, there are six possible results. Hence there are thirty-six possible states of the pair of dice [...]. Now in doing these school exercises in probability, we were introduced at a tender age to a set of (miniature) 'possible worlds'. The thirty-six possible states, *as long as we fictively ignore everything about the world except the two dice and what they show.* (Kripke 1980: 16, emphasis added)

Comparing miniature, or small, worlds to Lewisian big possible worlds, Kripke continues:

> The miniature worlds are tightly controlled, both as to the objects involved (two dice), the relevant properties (number on face showing) and (thus) the relevant idea of possibility. [Big Lewisian] 'possible worlds' are total 'ways the world might have been', or states or histories of the *entire* world. (Kripke 1980: 18)

Instead, small worlds do not match the actual and the counterfactual course of the world in a complete way.[13] Rather, small worlds are strongly restricted models of relevant factors.

> Perhaps such confusion [between small and big worlds] would have been less likely but for the terminological accident that 'possible worlds' rather than 'possible states', or 'histories', of the world, or 'counterfactual situations' had been used. Certainly they would have been avoided had philosophers adhered to the common practices of schoolchildren and probabilists. (Kripke 1980: 20)

The obvious analogy of the dice case and causal models is that the two dice and the possible outcomes of their throws correspond to variables and their possible values in a model. This fits the view of models in the debate on causation if we add laws (or structural equations). These laws describe what happens to the values of all variables when one variable has a certain value – they are, in this sense, analogue to the probabilities used in Kripke's example. In the debate on causation, a specific assignment of values to the variables in a model is regarded as a Kripkean small world. Causal model M (Chapter 2, pp. 44–5) is a triple $\langle U, V, L \rangle$ of a set of endogenous variables $V$, a set of exogenous variables $U$ and a set of laws L. For instance, according to Pearl (2000: 207, definition 7.1.8) a world is a particular realization or assignment of values to the variables of a model M.[14]

Do we have to believe that the objections given here are well-justified? Not necessarily. One might try to defend Lewisian semantics in its orthodox form. But let us suppose, for the sake of argument, that the interventionist objections are plausible to a considerable extent. In the recent debate on counterfactual theories of causation, several philosophers (Pearl 2000; Hüttemann 2004; Loewer 2007; Maudlin 2007; Leitgeb 2012) have employed a non-standard semantics. The general punch-line that I share with these philosophers is that, in order to evaluate counterfactuals, we have to refer directly to general statements (e.g., universal, probabilistic or ceteris paribus laws). Further, these philosophers defend a counterfactual theory of causation in Maudlin's (2004: 420) words – on a third factor which links causation and counterfactuals: this link consists in generalizations. The next three sections consider three alternative accounts of the meaning of counterfactuals for interventionists:

- The first approach is a modification of Lewisian semantics that deviates from the Lewisian standard with respect to selecting the closest worlds, because it uses a different heuristic to measure the closeness of worlds. The modified account also accepts the objection against big worlds.
- The second approach is more radical: relying on a meta-linguistic account, one accepts all of the objections and avoids possible worlds semantics completely.
- The third approach consists in the suppositional theory of counterfactuals. Contrary to the previous approaches, the suppositional account denies that counterfactuals have truth conditions. Instead, the meaning consists in pragmatic assertability conditions in terms of the Ramsey test.

## Modifying possible world semantics

Of course, the objections against Lewisian semantics are not knock-down arguments for a Lewisian. So, the most conservative approach for an interventionist would be to simply modify the Lewisian semantics. A sophisticated Lewisian may: (1) accept the objection against big worlds and convert to viewing worlds as small worlds, and (2) may counter the objections against similarity by spelling out 'the closest antecedent-worlds' differently than in terms of Lewis's similarity heuristic. Following this strategy preserves Lewisian semantics, but the entities that are ordered with respect to their similarity concerning facts and laws are small worlds, and the similarity ordering is given in a non-arbitrary and precise way. Let us identify this as the *conservative* approach to semantics.[15]

What is the alternative way to determine the closest antecedent-worlds? Consider the counterfactual 'if $X = x$ were the case, then $Y = y$ would be the case'. And bear in mind that we call a specific assignment of values to the variables in a model a (small) possible world. According to Woodward and Hitchcock's (2003: 13f.) tentative approach to an interventionist semantics, first, one might take the closest antecedent-worlds to be the worlds in which an intervention $I = i$ is carried out on the cause variable $X$ such that $X = x$. Second, these worlds are made up of the same variables as the actual[16] world (i.e. these worlds are assignments of variables to the same causal model M, apart from the intervention variable I), and other causes of $X$ and $Y$ that are not on a path from $X$ to $Y$ are held fixed. In other words, in the closest possible worlds the off-path causes of $X$ and $Y$ are set on a value that does not change $Y$ if $X = x$ (this is the 'redundancy range' for the causes of $Y$, as presented in Chapter 2, pp. 37–8). Third, if $Y = y$ is true in the closest $X = x$-worlds, then the counterfactual is true. Let us recapitulate on these three steps by summing up an interventionist semantics in a generalized form:

> *Interventionist possible worlds semantics*: A counterfactual if $\phi$ *were the case, then $\psi$ would be the case*[17] is (non-vacuously) true at a small world w (relative to a causal model M) if and only if $\psi$ is true in the closest $\phi$-worlds.

Most importantly, interventionists can maintain that the closeness relation is a weak ordering relation. That is, (a) a (small) world w might be equally close to two (small) worlds u and v, and (b) any two (small) worlds are comparable in their closeness. At this point, it is helpful to follow an idea by Judea Pearl: Pearl (2000: 241) proposes an interpretation of the closeness relation in terms of interventions by assuming 'an obvious distance measure among worlds, $d(w, w')$, given by the minimal number of local interventions needed for transforming w into $w'$'. Building on this idea, I propose that interventionists could measure closeness in terms of interventions by the following heuristic:

> Interventionist heuristic to measure closeness: A world u is among the closest $\phi$-worlds of w if:
> 1. there is an intervention such that $\phi$ obtains in u;

2. u and v consist of the same variables (except for the intervention variable) and the same functional relations (i.e. the same laws) obtain between the variables;

3. all the variables that are not on a directed path from the antecedent variable to the consequent variable take values in the redundancy range.

Worlds are more distant than the closest worlds to the degree that (a) they differ in interventions, (b) they instantiate different laws, and (c) they contain variables that do not take values in the redundancy range. What is the gain of this interventionist version of possible worlds semantics? On the one hand, this interventionist semantics agrees with Lewisian semantics on the general framework of possible worlds semantics. On the other hand, interventionist semantics diverges from Lewisian semantics in two important respects: first, worlds are small worlds, and, second, closeness of worlds is measure by a heuristic that builds on interventions.

Can this semantics handle Woodward's two counter-examples to the orthodox Lewisian semantics? Consider the car-scenario, first. The counterfactual 'if you had not been driving in the left lane (that is, if you had been driving in the right lane), then you would not have been late' is supposed to be evaluated as true. Woodward argues that the counterfactual comes out as false because of Lewis's similarity heuristic. The interventionist version of possible worlds semantics does not rely on this similarity heuristic. In order to evaluate the counterfactual, the position of the driver's car is represented by variable $D$ (ranging over a set of possible values {left; right}, a variable $T$ represents the time when the driver arrives at his destination ($T$ ranges over {late; on time}), and the behaviour of other cars has to be modelled by additional, off-path variables $W_1, ..., W_n$, each ranging over {left; right; not on this highway}. $W_1, ..., W_n$ have to be set to values in the redundancy range with respect to $D$ and $T$: for example, $W_1, ..., W_n$ take the value *left* or *not on this highway*. If one chooses worlds of this kind, then the counterfactual comes out as true.

Let us now turn to Woodward's second counter-example to orthodox Lewisian semantics. This counter-example is supposed to show that Lewis's similarity heuristic picks out the wrong worlds in order to describe a specific type of causal structure. Suppose that, at $t_0$, $X$ directly and independently causes the occurrence of each of $E_1, ..., E_n$ at $t_1$ and of $Y$ at $t_2$. According to Woodward, the counterfactual 'if $E_1, ..., E_n$ had not occurred, then Y would not have occurred' should be evaluated as false, because, in this situation at hand, $E_1$, ..., $E_n$ are described as irrelevant of for the occurrence of $Y$. According to the interventionist version of possible worlds semantics, the counterfactual is, indeed, evaluated as false in the closest antecedent-worlds, because if we hold $X$ at its actual value and we hold fixed $E_1$, ..., $E_n$, then $Y$ still occurs. So, the interventionist version of possible worlds semantics provides the correct result. For brevity's sake, let me simply suggest that analogous strategies to deal with Woodward's two counter-examples to Lewis's orthodox semantics are available for the interventionist versions of Goodmanian semantics and the suppositional theory, as will be discussed. Analogous strategies are available because the two alternative accounts of the meaning of counterfactual conditionals also use the crucial idea of a redundancy range condition.

Obviously, this interventionist semantics is not a reductive account of the meaning of counterfactuals, because it uses causal (and further counterfactual) information about interventions and other causes in order to evaluate a target counterfactual. This non-reductive approach is at odds with Lewis's reductive overall project in which his possible worlds semantics is embedded. However, possible worlds semantics is in no way committed to a reductive approach. For instance, already Stalnaker's possible worlds semantics is less ambitious. Stalnaker (1968: 33f.) does not expect there to be an informative and reductive analysis of 'an A-world which differs minimally from the actual world' which could be specified independently of judgements about what else would have been true if A were true. Unlike Lewis, Stalnaker does not seek a genuine analysis of counterfactuals in terms that do not presuppose counterfactual and causal relations.

The interventionist version of Lewisian semantics differs from a prima facie similar approach advocated by Jonathan Schaffer (2004b). Schaffer proposes to add causal independence conditions to Lewis's original similarity heuristic in order to determine the closest possible A-worlds in the following way:

1. It is of the first importance to avoid big miracles.
2. It is of the second importance to maximize the region of perfect match, *from those regions causally independent of whether or not the antecedent obtains.*
3. It is of the third importance to avoid small miracles.
4. It is of the fourth importance to maximize the spatiotemporal region of approximate match, *from those regions causally independent of whether or not the antecedent obtains* (Schaffer 2004b: 305).

Schaffer seems to count only those 'regions' that are not (direct or indirect) effects of the antecedent. Consequently, this characterization of relevant regions also applies to factors that are causally irrelevant to the cause and effect in question. Interventionists agree that focusing on the effects of the cause variable does not help to determine the closest worlds. But they are more specific on what does count to select the closest worlds: the causes of the antecedent (the cause) and the consequent (the effect). Nevertheless, at least the first and the second of Schaffer's criteria seem to have direct analogues in the interventionist framework: the first criterion of avoiding big miracles translates into 'select those worlds where the same laws (or generalizations) hold as in the world w'. The second criterion has an analogue in 'select those worlds at which (a) the causes of $X$ and $Y$ are held constant at values in the redundancy range, and (b) other variables have the same values as in the world w'. The interventionist heuristic to measure closeness avoids the vague notions of big and small miracles. This is an advantage over Schaffer's revision of Lewis's similarity heuristic.

The pay-off for the conservative approach is that Lewisian semantics is preserved. But one may feel that it would be unusual to accept Lewisian semantics if one defends an interventionist theory of causation? Is it not the case that, if one adopts Lewisian semantics, the interventionist theory of causation collapses into Lewis's counterfactual theory? Answer: Yes, there seems to be a prima facie problem for an interventionist who wants to be a sophisticated Lewisian. The choice of formal semantics does not mark a difference between a Lewisian and an interventionist theory of causation. A blatant difference that divides Lewis and Woodward is methodological: Lewis defines causation in entirely non-causal terms by evaluating counterfactuals in non-causal terms, but interventionists do not. As opposed to this reductive strategy, interventionists avoid the threat of collapse because their modification of the closeness relation is non-reductive; that is, it refers to additional causal information. Whether one accepts non-reductive truth conditions for counterfactual and causal statements is another, separable line of demarcation between the orthodox Lewisian (or Humean) camp and the interventionist camp – not merely the choice of semantics.[18] I believe that arguments for and against non-reductive explications can be detached from the question of which semantics one chooses (the tenability of non-reductive explications is discussed in Chapter 7).

Further, it is no longer unclear whether laws play a role in interventionist semantics. Following Lewis and Schaffer, the modified version of possible worlds semantics is in agreement with the assumption that the most important criterion for selecting the closest possible worlds are laws (or generalizations). A clear advantage of the interventionist heuristic to measure closeness is that interventionists can calculate whether a counterfactual is true relative to a causal model. Hence, the interventionist heuristic avoids the vagueness of Lewis's and Schaffer's similarity heuristics.

## Modified Goodmanian semantics

The second option for interventionists is the meta-linguistic approach to counterfactuals. In his influential book *Fact, Fiction and Forecast*, Goodman (1983: 8f.)[19] famously argues for the view that counterfactuals are condensed or hidden arguments. For instance, consider Goodman's classic evaluation of the counterfactual 'Had the match *m* been struck, then it would have lit'. This counterfactual is true iff the antecedent (i.e. the proposition that the match *m* is struck), certain boundary conditions (e.g. the match is dry, properly made, sufficient oxygen is present) and a set L of laws of nature deductively imply the consequent (i.e. the singular proposition that the match *m* lights). In general, these arguments contain a set of law statements and singular statements (most importantly singular statements describing a counterfactual situation expressed by the antecedent proposition of the counterfactual conditional) as its premises. The conclusion, which follows from these premises, is another singular statement. If the counterfactual is true, the conclusion of the argument is the consequent proposition of the counterfactual conditional, otherwise the conditional is false.

Two of the most elaborated approaches by Judea Pearl and Tim Maudlin to the semantics for counterfactual conditionals that seem to revive the meta-linguistic account (similarly, cf. Kvart 1992, 2001) are Pearl's structural model semantics and Maudlin's modest proposal of a three-step semantics.

### Pearl's structural model semantics

A plausible way to understand Judea Pearl's semantics is to reconstruct it as a meta-linguistic account. Pearl provides a semantics for counterfactual conditionals straightforwardly with respect to causal models. Recall that a causal model $M$ is a triple $\langle U, V, F \rangle$ consisting of: a set U of exogenous variables, a set V of endogenous variables and a set L of structural equations or laws.[20] Pearl describes the semantic evaluation of a counterfactual in the following way: 'The sentence "$Y$ would be $y$ had $X$ been $x$" [...] is interpreted to mean "the solution for $Y$ in $M_x$ is equal to $y$ under the current conditions $U = u$''' (Pearl 2000: 100). In this formulation, '$M_x$' denotes those structural equations (laws) in which $X$ has taken the value $x$. In his book *Causality*, Pearl is even more clear on the issue that evaluating counterfactuals crucially depends on general statements; namely, on the structural equations of a model $M$: 'Let $X$ and $Y$ be two subsets of variables in **V**. The counterfactual sentence "The value that $Y$ would have obtained, had $X$ been $x$" is interpreted as denoting the *potential response* $Y_x(u)$' (Pearl 2000: 204, definition 7.1.5). A *potential response* $Y_x(u)$ of $Y$ to setting the value of $X$ to $x$ (given that the exogenous variables $U$ have the value $u$) is defined as '*the solution for $Y$ of the set of equations $F_x$*' (Pearl 2000: 204, definition 7.1.4). As explained above with respect to the expression '$M_x$', Pearl's notation '$F_x$' denotes the fact that $X$ has taken the value $x$ in those structural equations, which variable $X$ plays a role in. According to Pearl (2000: 206), a counterfactual 'if A were the case, then B would be the case' is supposed to be semantically evaluated, roughly, in the following three steps:

- *Step 1 – Abduction*: Update the boundary conditions by evidence $e$.
- *Step 2 – Action*: Modify the causal model $M$ by action do(A), where A is the antecedent of the counterfactual, to obtain the submodel $M_A$.
- *Step 3 – Prediction*: Use the modified model to compute the probability of B, the consequent of the counterfactual.

Based on this idea, Pearl develops a formal semantics for evaluating counterfactuals in deterministic and probabilistic models. The formal details of this semantics do not matter a great deal for the concerns of this book. Rather, the upshot of Pearl's semantics is that one needs to rely on models, and especially on structural equations, in order to evaluate counterfactual conditionals. One evaluates counterfactuals by inferring the value of the consequent variable from three kinds of premises: (a) start with a causal model (that is supported by evidence), (b) then take one premise that reports setting the antecedent variable at some possible value while other variables are held fixed (this is expressed by the 'actions' and the 'do-operator', Pearl's notion of an intervention), and (c) one premise that states information about the laws in the model (the laws are crucial for computing the probability of the consequent).

However, it is important to point out that Pearl does not use Woodward's notion of an intervention. To set a variable $X$ to a certain value (while other variables are held fixed) does not require an intervention that is an exogenous cause of $X$. Although Pearl sometimes uses the notion of a surgical intervention, he usually uses the term 'action' or the 'do-operator'. For instance, suppose that the antecedent of a counterfactual is $X = x$. According to Pearl's Step 2, we set $X$ to the value $x$ simply by constructing a model in which $X = x$ is true (and other variables are held fixed). As Pearl (2000: 223–5) explicitly underlines, in addition to interventions (or simply, as Pearl says, 'facts' that are results of 'local surgeries') one needs laws (or 'mechanisms', as Pearl sometimes alternatively denotes them) in order to evaluate counterfactuals. Yet, from an interventionist point of view, it nonetheless seems to be possible to use the stronger Woodwardian notion of an intervention in the framework of Pearl's structural model semantics.

I propose to reconstruct Pearl's semantics as a meta-linguistic semantics. Having this in mind, it is noteworthy to draw attention to the distinction between (a) Lewisian possible worlds semantics, (b) the Lewisian similarity heuristic, and (c) Lewis's logic for counterfactuals (cf. Lewis 1973a: ch. 6 on his V-logic). First, concerning Lewisian semantics, Pearl argues that his structural model semantics can be reconstructed as a possible worlds semantics. Second, concerning the Lewisian similarity heuristic, Pearl (2000: 239f.) denies that his approach is forced to rely on a similarity heuristic. He claims instead that the semantic evaluation of counterfactuals 'rest[s] directly on the mechanisms (or "laws", to be fancy) [...] Thus, similarities and priorities [...] are not basic to the analysis.' Third, and perhaps most importantly, Pearl (2000: 240–3, Pearl and Galles 1996, 1998) argues that his axioms of the logic of counterfactuals (i.e. the axioms of effectiveness, composition, and reversibility) are equivalent to the axioms of Lewis's logic for counterfactuals (cf. Lewis 1973a: 123, 132).

### Maudlin's modest proposal of a three-step semantics

In the tradition of Goodmanian semantics, Tim Maudlin's semantics for counterfactuals also builds straightforwardly on laws. His proposal is less technical than Pearl's structural model semantics and it does not necessarily require the framework of causal modelling. According to Maudlin, laws describe how the behaviour of systems evolves over time.[21] This characterizes the fundamental laws of physics as well as the laws in the special sciences. The latter describe how a system would behave through time 'under normal conditions' or 'if nothing interferes' (Maudlin 2007: 12–14). Being non-universal does not prohibit those special science generalizations to be like laws, insofar as they are accepted to support counterfactuals and to provide explanations in scientific practice (non-universal laws will be further discussed in Chapter 5). One central case of a counterfactual that Maudlin discusses is 'if the bomb on Hiroshima had contained titanium instead of uranium, then it would not have exploded'. Maudlin offers a three-step semantics to determine the truth-value of the counterfactual:

> *Step 1*: Choose a Cauchy surface that cuts through the actual world and that intersects the bomb about the time it was released from the plane. All physical magnitudes take some value on this surface.
> *Step 2*: Construct a Cauchy surface just like the one in Step 1 save that the physical magnitudes are changed in this way: uranium is replaced by titanium in the bomb.
> *Step 3*: Allow the laws to operate on this Cauchy surface with the new boundary values generating a new model. In that model, the bomb does not explode.
> *Ergo (if we have got the laws right, etc.) the counterfactual is true* (Maudlin 2007: 23, emphasis added).[22]

A Cauchy surface is a mathematical representation of an entire (small) world state. Such a surface describes the boundary conditions that are required in the semantics (Maudlin 2007: 18). Abstracting from the concrete example, the general idea of the three steps appears to be: in a first step, determine the actual boundary conditions (i.e. the values of the variables, or 'magnitudes') in question. In a second step, change the value of the variable to which the antecedent of the counterfactual refers to the value that the antecedent assigns to it. In a third step, check whether the consequent of the counterfactual follows, or can be calculated, from the new boundary conditions (as constructed in the second step) and a law that applies to the boundary condition. If the consequent does, indeed, follow, the counterfactual is true; if it does not follow, then the counterfactual is false (in Maudlin's bomb example it does not follow). As Maudlin underlines, the three-step semantics applies to universal as well as to non-universal laws in the special sciences in general, and, as he stresses, in the social sciences (cf. Maudlin 2007: 45. Chapter 5 expands on this point).

So, what is the Goodmanian spirit that is common to Pearl's and Maudlin's approaches? I take the heart of Goodmanian semantics to be the idea that counterfactuals are to be reconstructed as arguments. These arguments are supposed to have the following form: they contain nomological and singular statements – including the antecedent of the counterfactual and statements about other relevant factors – as premises and the consequent of the counterfactual as their conclusion. If such an argument is sound, then the counterfactual is true. I follow Pearl and

Maudlin in determining the truth-value of a counterfactual relative to a set of laws and a set of parameters (i.e. relative to a model). I think it is a fair interpretation of Pearl and Maudlin[23] to ascribe the claim to them that the realization of the antecedent can be conceived as an adjustment of a mathematical model that need not satisfy the formal constraints of a Woodwardian intervention. However, I think that interventionists can use the general framework for their own purposes. Is there a concise way to express the essential Goodmanian idea of the semantics by Pearl and Maudlin? And can we make it useful for the project of an interventionist semantics? Here is an attempt to do so:

> Interventionist modified Goodmanian semantics. *If $\phi$ were the case, then $\psi$ would be the case* is true relative to a causal model M= $\langle$U, V, L$\rangle$ iff the consequent of the conditional $\psi$ follows from a set of statements: (a) a statement expressing that an intervention exists such that the antecedent of the conditional $\phi$ is true, (b) the statement $U_i = u_i$ (expressing that we assume some variables as exogenous ones), (c) the statement that other variables in V are in the redundancy range (or, more generally, a ceteris paribus clause is used to represent that these variables are being held constant), and (d) the laws $L$.

Here, U is a set of exogenous variables ($U_i = u_i$ is a statement that expresses the proposition that the exogenous variables are set to certain values), V is a set of endogenous variables, and L a set of laws.

Let us see how this semantics works for a textbook example from economics (this example is used for the same purpose by Pearl 2000: 215–17). The law of supply states: if the supply of a commodity increases (decreases) while the demand for it stays the same, then the price decreases (increases) (cf. Roberts 2004: 159; also Kincaid 2004: 177). We can reconstruct this law as being part of a causal model consisting of the laws of supply and demand (as set of laws L), and variables for supply and price (as set of endogenous variables V) and for demand (as an exogenous variable U) of a good. So, is the counterfactual 'if the supply of a commodity increased, then the price would decrease' true relative to the model? First, we suppose that the law of supply and demand is true (nomological premise). Second, we suppose that the supply increases (endogenous variable) and that the demand stays the same (exogenous variable), as the antecedent of the counterfactual says. This second step might be described in terms of interventions on the supply and demand. Third, from these premises it follows that the price increases. Therefore, the counterfactual is true relative to a model consisting of the laws of supply and demand (as set of laws L), and variables for supply and price (as set of endogenous variables V) and for demand (as set of exogenous variables U) of a good.

The proposal of an interventionist version of Goodmanian semantics is a modified version of Goodman's original semantics in two respects.

First, the semantics I propose as an option for interventionists differs from Goodman's original approach in reductive spirit. Goodman (1983: 20) originally intended to evaluate counterfactuals without employing further causal information. This reductive premise has a price for Goodman: he struggles to determine factors that are relevant for the truth of the counterfactual besides the singular statements φ and ψ, and the set of general statements L. Goodman (1983: 8) diagnoses the difficulty of determining further relevant conditions to be one of the two 'major problems' for a semantics of counterfactuals.[24] In a sense, the major problem of determining relevant conditions does not arise from an interventionist point of view. First, it does not arise because the interventionist approach does not aspire to be a reductive explication. Second, it does not arise, because (some) relevant conditions are already part of the causal model as other causes. However, one may ask: 'What about those causes that are not part of the model?' A modest reply by interventionists could be: 'Our semantics is model-dependent'; that is, a counterfactual is true relative to a set of variables and a set of laws. And more – something non-relative – one cannot say. (In Chapter 5 (p. 140), I attempt to dissolve this model-relativity by distinguishing an epistemological and a metaphysical reading of quasi-Newtonian laws.)

Second, according to interventionist modified Goodmanian semantics, the consequent of the counterfactual follows from a set of premises, if the counterfactual in question is true. For Goodman, this inference has to be deductively valid. One reason to assume deductively valid rules of inference is the idea that laws are necessarily (but not sufficiently) universally quantified statements. One can assume deductive entailment because, given the universal and complete laws, a complete determination of initial conditions allows the inference of the consequent of a true counterfactual. The trouble with this idea is, as Goodman himself notes, to determine the set of all relevant initial conditions. What if one gives up the determination of all relevant conditions? Another option would be to acknowledge that counterfactuals do not allow the strengthening of an antecedent. In the recent debate, this is the received view concerning a logical property of counterfactuals (cf. Lewis 1973a: 31, Bennett 2003: 159–61, Sider 2010: s. 8.1.1).

In other words, it seems reasonable to say that this inference (from laws and factual information) cannot be monotonic. An inference is monotonic iff, for all sets of premises, adding an arbitrary set of new premises to a valid argument does not affect its validity. Notice that the non-monotonicity of inferences is a logical property of counterfactual reasoning that also plays an important role for Lewis's view (Lewis 1973a: 31f.). The non-monotonic character of inference becomes obvious when we take a look at the special sciences such as biology and economics: in these disciplines, general statements cannot be formally understood as universally quantified sentences. Rather, they are probabilistic generalizations or ceteris paribus laws (see Chapter 5). The inference from a set of premises (including a set probabilistic or ceteris paribus laws $L$ and a singular statement $S$) to a conclusion $C$ is not truth-preserving, if a disturbing factor $H$ occurs.[25] In other words, it does not follow from the original set of premises (i.e. the laws $L$ and $S$) plus the new premise (asserting that $H$ occurs) that $C$ is the case. Nevertheless, this result does not imply that inferences, involving probabilistic or ceteris paribus laws, are an arbitrary matter. Quite the contrary, recent developments in non-montonic logic provide the formal tool to show how reasoning with probabilistic or ceteris paribus laws is possible (for various attempts to provide a precise understanding of 'logically follows' in the modified version of Goodmanian semantics cf. Kvart 1992, 2001; Gabbay et al. 1994; Schurz 1997, 1998; Leitgeb 2004). It seems to be a neat side-effect of these considerations with respect to laws in the special sciences that we can explain why the strengthening of an antecedent does not hold for counterfactuals, because the laws that are employed in order to evaluate a counterfactual are typically not universal and complete, as Goodman (and many others) assumed.

So, what is the pay-off of interventionist modified Goodmanian semantics from an interventionist point of view? The semantics does not rely on a similarity relation. Hence, it avoids the objections against similarity. Since a counterfactual is assigned a truth-value relative to a model, the semantics suggests (and, at least, does not preclude) the interpretation of worlds as small worlds. Hence, the interventionist version of Goodmanian semantics avoids the objection against big worlds. If interventionists adopt Goodmanian semantics, then this feature of interventionism does clearly distinguish interventionism from the Lewisian version of a counterfactual theory of causation. Additionally, the interventionist brand of Goodmanian semantics promises to deal with the two counter-examples to Lewis's original semantics (see

'Objection against Similarity I', pp. 82–5), in the same way as the modified possible worlds semantics. For reasons of space, this matter will not be discussed in detail. We may conclude that the interventionist modified Goodmanian semantics for counterfactuals seems to be a candidate for supporting an interventionist theory of causation.

## The suppositional view: the pragmatic meaning of counterfactuals

A well-established third option for interventionists to account for the meaning of a counterfactual – known as the suppositional theory – goes back to an idea by Frank P. Ramsey that is often referred to as 'the Ramsey test' for conditionals. Ramsey claims:

> If two people are arguing 'If p will q?' and are both in doubt as to p, they are adding p hypothetically to their stock of knowledge and arguing on that basis about q [...]. We can say that they are fixing their degrees of belief in q given p. (Ramsey 1929: 247)

The Ramsey test is a test for the correctness of asserting a conditional, and aims to provide assertability conditions. That is, the Ramsey test is a test for whether it is correct to assert a conditional: if adding the antecedent 'hypothetically' to one's knowledge, then one also believes that the consequent is true. Hypothetically adding the antecedent to one's beliefs is an epistemic operation. It is an assumption or a supposition. Note that this suppositional approach to the meaning of counterfactual conditionals differs from the truth-conditional approaches such as Lewisian and Goodmanian semantics. According to the suppositional theory, conditionals do not have truth conditions (and they do not have a truth-value). Instead, the meaning of a conditional is not semantic but pragmatic, in the sense that, in communication, it is correct to assert a conditional to the extent that an epistemic agent accepts the consequent of a conditional under the supposition of the antecedent. Let us call this the 'suppositional theory' (cf. Edgington 1995, 2008).

The basic idea of the Ramsey test was historically first used to determine the assertability conditions of indicative conditionals by Adams (1975). Adams analyses the degree of assertability of an indicative conditional in terms of subjective probability: the degree of assertability of 'if $p$, then will $q$' equals the subjective probability of $q$ given $p$. Although this fact is often ignored, Adams (1975: ch. 4) also argues that the suppositional theory can be applied to counterfactuals in a slightly modified way. In particular, Skyrms has developed the most sophisticated account of Adams's original idea.[26] According to Skyrms (1980a: 261; 1984: 100f; 1994: 13–15), one determines the pragmatic meaning of a counterfactual in terms of the degree of assertability:

> The degree of assertability of the counterfactual 'if it were the case that p, then it would be the case that $q$' equals the subjectively expected objective conditional probability of $q$ given $p$.

So, if the expected objective conditional probability of $q$ given $p$ is sufficiently high, then any rational epistemic agent is supposed to assert the counterfactual 'if it were the case that $p$, then it would be the case that $q$'. Otherwise – that is, if the expected objective conditional probability of $q$ given $p$ is low – a rational epistemic agent is supposed not to add $q$ to his or her stock of beliefs. Skyrms's account is a formalized version of the Ramsey test, because the assertability of a counterfactual is evaluated by the expected objective probability of $q$ *under the supposition that p*. In other words, the principal idea of Skyrms's account is that the antecedent – understood as a 'contrary-to-fact' supposition (cf. Joyce 1999) – picks out an epistemically possible world w where (at least) the objective probabilities of $q$ given $p$ are similar to the actual world. Does the suppositional account – in analogy to possible worlds semantics and Goodmanian semantics – rely on laws? Adams, Skyrms and Edgington are not explicitly concerned with this issue. Nevertheless, it seems to be obvious that what an epistemic agent expects certain objective conditional probabilities to be depends crucially on which laws and causal relations are instantiated in the agent's epistemically possible worlds (Leitgeb 2010, s. 2).

Finally, let us ask whether the suppositional theory is a viable option for interventionists. At first glance, this seems to be the case: if the degree of assertability of the counterfactual 'if it were the case that $p$, then it would be that case that $q$' equals the subjectively expected objective conditional probability of $q$ given $p$, then interventionists may add an amendment of the following kind: the supposition that $p$ is the outcome of an intervention, or, alternatively, an epistemic agent supposes that there is an intervention on the antecedent variable of the conditional such that $p$ is the case. This amendment seems to be perfectly compatible with the suppositional approaches. Furthermore, if one takes Woodward's 'pragmatic considerations' (Woodward 2004: 45) seriously, then he might be understood as referring merely to the pragmatic meaning of counterfactuals. Pearl's evaluation of counter-factuals (pp. 92–4) might be also interpreted according to the suppositional view, because truth conditions are not always mentioned and the first step in Pearl's three-step recipe (updating the antecedent variable) might be regarded as a supposition (cf. Pearl 1994: 71 refers to Skyrms's work). Regardless of whether this interpretation is possible and ultimately convincing, adopting the suppositional theory would have important consequences for interventionists: the suppositional account contradicts Woodward's claims that counterfactuals have objective truth-values and that their meaning can be accounted for by giving truth conditions. Obviously, interventionists could simply give up their semantic project and, instead, adopt the suppositional account concerning merely the pragmatic meaning of counterfactuals. But following this strategy would commit them to the claim that causal statements also lack truth conditions. Instead of being true or false, causal statements have a degree of assertability. It is questionable whether this is acceptable to many philosophers in the debate, and to interventionists in particular.[27] I take Woodward's (1992: 215, fn. 11; 2003: 299–303) repeated critique of Skyrms's (1980b) epistemic-pragmatic approach to causation and laws to be an indication that abandoning objective truth conditions of causal claims is not an acceptable and welcome option for Woodward (which is probably the case for other interventionists).

## Conclusion

This chapter argued that interventionists should provide an account of the meaning of counterfactuals, because they claim that counterfactuals have truth conditions without revealing what exactly these truth conditions are. The natural response to this situation is to provide such an account of meaning. Since interventionists argue against the orthodox Lewisian semantics for counterfactuals, three alternative standard approaches were presented: (a) an interventionist modified possible worlds semantics, (b) an interventionist modified Goodmanian semantics, and (c) a suppositional theory of pragmatic meaning. These semantics successfully cope with two counter-examples that were raised by Woodward against Lewis's original semantics. The crucial question revolves around which semantics is the right one. There is no clear and absolute answer to this question. And it is important to point out that all three approaches are viable options for interventionists. Probably, the best one can do is to use the following interventionist claims about

counterfactuals and causation as a conditional measure of the success
for each of the three interventionist semantics:

- causal statements and counterfactual claims have meaning;
- the meaning of causal and counterfactual statements is determined
  semantically by truth conditions;
- in order to state truth conditions, one needs to invoke causal
  notions.

First, an *interventionist modified possible worlds semantics* satisfies all three
of these claims: (a) counterfactuals and causal statements have meaning; (b) this semantics preserves the claim that the meaning of counterfactuals (and causal statements) is fixed by truth conditions; and (c) this
semantics fits a non-reductive explication of causal notions, because
it explicitly refers to causal information. A further reason to pick possible worlds semantics consists in the fact that it is acknowledged to
be a powerful formal tool in many philosophical debates. Second, the
*interventionist modified Goodmanian semantics* also conforms to the three
claims. Third, adopting the *interventionist version of the suppositional view*
would force interventionists to reject the claim that causal statements
and counterfactuals have a semantic meaning that can be determined
by stating truth conditions. A suppositionalist interventionist would
have to assert that causal statements and counterfactuals exclusively
have a pragmatic meaning that can be determined in terms of acceptability conditions. Even so, adopting a suppositional view is by no
means excluded by the interventionist definitions of causal notions. It
seems to be perfectly compatible with these definitions. Interventionists
would have to be willing to pay a price for accepting the suppositional
view: interventionists would have to make up their minds about the
metaphysics of causation. The suppositional view would suggest an
epistemic nature of causation as opposed to causation as a feature of the
(physical) objective, non-epistemic world (cf. Williamson 2005: ch. 9;
Beebee 2007; Price 2007, for leading work on epistemic causation).

  Again, all of these options are viable for advocates of an interventionist theory of causation. So, where does this leave us? The fact that one
cannot properly prefer one semantics over another from an interventionist point of view suggests that there is a gap or incompleteness in
the interventionist theories of causation. The gap or incompleteness
is yet to be overcome. However, this result should not worry interventionists too much. Rather, having three alternative accounts of the
meaning of counterfactuals should encourage interventionists to make

their theory of causation more precise. A promising way to achieve this precision might consist in exploring the consequences of adopting each of the accounts of meaning for counterfactuals.

In this chapter, I took the notion of a possible intervention seriously. For the sake of argument, I attempted to meet a challenge for interventionists: to provide an account of truth conditions (and, more generally speaking, an account of the meaning) of interventionist counterfactuals. However, Chapters 4, 5, and 6 will argue that the notion of a possible intervention and the versions of interventionist semantics for counterfactuals relying on this notion are problematic.

# 4
## Getting Rid of Interventions

### Introduction

Chapters 2 and 3 stressed the virtues of Woodward's interventionist theory of causation; Chapter 2 diagnosed that the interventionist theory conforms to, or at least promises to conform to, the naturalist criterion and the distinction criterion of adequacy. Consequently, the interventionist theory of causation, prima facie, appears to be a candidate for an adequate explication of causation in the social sciences, and, hopefully, of causation in the special sciences in general. This is, at first glance, a great success for Woodward's interventionist theory. Chapter 3 drew attention to a problem for interventionists: they fail to provide an explicit account of the truth conditions of interventionist counterfactuals, the building blocks of the interventionist theory of causation. For the sake of argument, I argued in favour of the interventionist theory by proposing three different semantics for interventionist counterfactuals (i.e., the interventionist versions of possible worlds semantics, Goodmanian semantics, and the suppositional theory).

Chapter 4 begins a critique of the interventionist theory of causation (which is continued in Chapters 5 and 6). The main negative target consists in arguing that the central concept of the interventionist theory of causation – that is, the concept of a possible intervention – is deeply problematic. For this reason, it is propounded that the interventionist theory of causation is not tenable.

Despite the prima facie virtues of interventionist theories of causation, the key notion of the interventionist approach – the notion of a possible intervention – turns out to be deeply problematic. Recall the central character of interventions in Woodward's framework: Woodward's definitions of various causal notions imply that the truth of a causal statement 'X directly causes Y' requires the existence of an possible intervention on X. Another way to make the same point is that, if 'X directly causes Y' is true, then the following interventionist counterfactuals have to be true: 'if there were an intervention $I = i$ on X such that $X = x$, then $Y = y$ would be the case', and 'if there were an intervention $I = i^*$ on X such that $X = x^*$, then $Y = y^*$ would be the case' (with $i \neq i^*$, $x \neq x^*$, $y \neq y^*$). Woodward requires that interventions be merely in principle possible. Woodward interprets in principle possibility as logical possibility. Let us call this the 'modal character' of interventions.

This chapter will argue that the key notion of the interventionist approach – the notion of a possible intervention – turns out to be deeply problematic: in particular, that Woodward's notion of an intervention is problematic because of the modal character of possible interventions. There are two claims against Woodward to be argued here:

- *Either* merely logically possible interventions are dispensable for the semantic project of providing an account of the meaning of causal statements. If interventions are indeed dispensable, the interventionist theory collapses into (some sort of) a counterfactual theory of causation.[1] Thus, the interventionist theory is not tenable as a theory of causation in its own right.
- *Or*, if one maintains that merely logically possible interventions are indispensable, then interventions with this modal character lead to the fatal result that interventionist counterfactuals are evaluated inadequately. Consequently, interventionists offer an inadequate theory of causation.

Both of these claims are compatible with the view that interventions are important (and perhaps even indispensable) for the methodological task of discovering causal relations. The next section will briefly reconstruct Woodward's notion of an intervention and will be followed by a discussion of which kind of possibility is presupposed in Woodward's notion of possible interventions. A case will be presented that is supposed to show that it would be unwise to require that interventions must be physically possible. Woodward himself suggests this kind of counter-example against the requirement that interventions need to be physically possible. Reacting to the counter-example, Woodward rejects the claim that interventions have to be physically possible and requires interventions to be merely logically possible.

The chapter goes on to present two arguments against the adequacy of an account of causation framed in terms of logically possible interventions. The first argument demonstrates that merely logically possible interventions are superfluous and can be dispensed with for pursuing the semantic project of providing truth-conditions for causal claims. The second argument takes a different stance, leaving aside the dispensability arguments. It will be shown that relying on merely logically possible interventions conflicts with the standard approaches to the meaning of counterfactuals. If an existential claim about a merely logically possible intervention figures in the antecedent of an interventionist counterfactual, then this leads to the fatal result that this interventionist counterfactual is evaluated as false, although we would take these counterfactuals to be true. It will also be argued that the obvious strategy to choose from an interventionist point of view – that is, to adopt an interventionist semantics of counterfactuals – does not solve the problem. Even so, the counter-arguments presented will be softened with a reconciling stance on the situation: although we should not take interventions to be part of the truth conditions of causal claims, interventions can still be of heuristic value. The chapter concludes by stating that, when analysing causal concepts and stating the truth conditions of causal claims, we would do best to get rid of interventions.

## Woodward's concept of intervention

Woodward defines interventions in two steps: first, he defines an intervention variable, then he uses the notion of an intervention variable in order to define the notion of an intervention. In the reconstructed version of Woodward's definition of an intervention variable, $I$ is an intervention variable for $X$ relative to $Y$ iff the conjunction of the following conditions is satisfied:

1. $I$ is a cause of $X$.
2. There is at least some value of $I$ such that if $I$ takes this value, then $X$ depends only on $I$ and $X$ depends on no other variables; that is, $I$ is the only cause of $X$.
3. $I$ is not a direct cause of $Y$, and if $I$ is a cause of $Y$ then $I$ is an indirect cause of $Y$ via a causal path leading through $X$ and a – possibly empty – set of intermediate variables $Z_1, ..., Z_n$.
4. $I$ is probabilistically independent of other causes $W_1, ..., W_n$ of $Y$, which are not on a causal path leading from $X$ to $Y$.
5. $I$ does not alter the relationship between $Y$ and its causes $W_1, ..., W_n$, which are not on a causal path leading from $X$ to $Y$.

The notion of an intervention variable is, in turn, used to define the notion of an intervention: any value $i_i$ of an intervention variable $I$ (for $X$ relative to $Y$) is an *intervention on $X$* iff it is the case that the value of $X$ counterfactually depends on the fact that $I$ has the value $i_i$ (cf. Woodward 2003: 98). Among other features of interventions, Woodward stresses the modal character of interventions. Interventions are not required to be possible in the sense that they are feasible actions for human agents. Woodward's key idea consists in the assumption that it is in principle possible, or logically possible, to intervene. The main goal of this chapter is to investigate the modal character of interventions and the problems arising from it.

## A counter-example: why some interventions are not even physically possible

Recall that, according to Woodward, if $X$ is a direct cause of Y, then there is a possible intervention on $X$ that changes $Y$. One might wonder which kind of possibility is in play in the interventionist theory of causation. Could the required kind of possibility simply refer to the abilities of human agents to intervene? Woodward and other interventionists explicitly (and rightly, I think) deny that interventions have to be practically possible; that is, interventions are not required to be feasible actions for human beings. Woodward (2003: 103f., 123–7) argues against this anthropocentric concept of an intervention as defended by advocates of agency theories of causation (such as Menzies and Price, 1993). Woodward (2003: 123) objects that agency theories are committed to two problematic metaphysical claims: (a) the capacity of agents to intervene is 'a fundamental and irreducible feature of the world and not just a variety of causal interaction among others', and (b) an agency theory 'leads us toward an undesirable kind of anthropomorphism or subjectivism regarding causation'. As Woodward believes that causal claims have mind-independent truthmakers, he rejects practical possibility (or abilities of an agent to intervene) as the right kind of modality required for his concept of an intervention.

So, is it adequate to think that interventions are physically possible? Woodward addresses this question by presenting and discussing a counter-example against the assumption that physically possible interventions are necessary for determining the truth conditions of causal claims:

> Suppose that [...] [$X$]s *occur only spontaneously in the sense that they themselves have no causes.* There are no further factors [$I_1$, ..., $I_n$] that affect whether or not [$X$] happens, and this is a matter of physical

law. (I take this at least to be a logically coherent possibility.) Thus, it is *physically impossible* to carry out an intervention that changes whether [X] occurs. Nonetheless, it seems quite possible that the [X]s themselves might well have further effects [Y]. (Woodward 2003: 130, my emphasis added, alteration of notation by this author)

The general form of Woodward's counter-example is:

X is a cause of Y but there is no physically possible intervention $I = i$ on X.

I will distinguish two senses of 'spontaneously' in the quote by Woodward.

- There are cases in which there is *no* physically possible *deterministic* intervention *but* there *is* a physically possible *indeterministic* intervention on X; and
- There are cases in which there is *neither* a physically possible *deterministic* intervention *nor* a physically possible *indeterministic* intervention on X.

Two concrete examples from physics illustrate these cases:

No-deterministic-but-indeterministic intervention: Uranium decays 'only spontaneously' in the sense that there is no deterministic, physically possible way to manipulate uranium such that it surely decays at a time *t*. Nonetheless, the decay of uranium causes a flash on a screen (as a measurement in an experimental set-up). Yet, one can raise the probability that the uranium atom decays (e.g. by increasing the energy of the nucleus of the uranium atom). In other words, there is a physically possible indeterministic intervention on uranium decay.

Neither-deterministic-nor-indeterministic intervention: Tim Maudlin (2002: 149f.) presents the Big Bang as a case where there is neither a deterministic physically possible way to intervene, nor an indeterministic way to intervene. In other words, according to Maudlin's example, there is no physically possible indeterministic cause of the Big Bang that raises or lowers the probability of the occurrence of the Big Bang. Nonetheless, one would like to maintain that the Big Bang – if anything – has a plentitude of direct and indirect effects.

The Big Bang case will be described in detail later in this chapter. However, the arguments offered do not depend on the Big Bang case. There are other 'less cosmological' examples of the neither-deterministic-nor-indeterministic intervention case. Although focus will mainly be on the Big Bang, let me briefly present two other examples.

*Physical constants*: Many people intuitively believe that physical constants have a causal influence on the behaviour of physical objects. For instance, the gravitational constant, the Planck constant, and the constant representing the speed of light can be taken to be causes. Even so, it is physically impossible to intervene on a physical constant because changing the constants amounts to changing the physical laws in a non-local way. Changing the laws creates physically impossible worlds.[2] Hence, we have another example of the neither-deterministic-nor-indeterministic intervention case: *if* we assume that physical constants count as causes,[3] then there is no physically possible intervention on a physical constant such that the value of the constant changes.

*Norton's dome*: John Norton (2007: 22–8) argues for a case of uncaused events that is compatible with Newtonian mechanics. Norton imagines a symmetrically shaped dome that is located in a downward-directed gravitational field. A mass is at rest on top of the dome, on the apex. Figure 4.1 represents the dome scenario.

Norton claims that the following scenario is consistent with Newtonian mechanics:

> It is a mass at rest in a physical environment that is completely unchanging for an arbitrary amount of time – a day, a month, an eon. Then without any external intervention or any change in the physical environment, the mass spontaneously moves off in an arbitrary direction with the theory supplying no probabilities for the time or direction of the motion (Norton, 2007: 22f.).

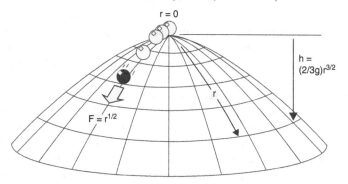

*Figure 4.1*  Norton's dome scenario
*Source*: Norton (2007: 23).

Norton interprets the acceleration of the mass as an uncaused event (cf. Norton, 2007: 24). Let us observe Norton's dome scenario thus: suppose that the acceleration of the mass has an effect $e$ (for instance, the mass hits a little bell that is located at the base of the dome).[4] Further suppose that we posit very restricted worlds such that the worlds only contain the mass, the dome, the gravitational field, and the effect $e$. In these restricted dome-worlds, there exists no physically possible event that may play the role of an intervention on the mass, because all that exists in the restricted dome-worlds is the mass, the dome, the gravitational field, and the effect $e$. In other words, restricted dome-worlds conform to Norton's initial description of the dome scenario – the idea that the 'physical environment that is completely unchanging for an arbitrary amount of time' – in an extreme way through the isolation of the dome. Hence, in these restricted dome-worlds it is the case that (a) the uncaused acceleration of the mass itself is a cause of event $e$, and (b) a physically possible intervention does not exist. In other words, we have a case of causation without the physical possibility of intervening. Therefore, if the restricted version of Norton's dome scenario is correct, then we have another example for the neither-deterministic-nor-indeterministic intervention case.

The neither-deterministic-nor-indeterministic intervention case is stronger than the no-deterministic-but-indeterministic intervention case, because Woodward can easily admit that interventions – and causation in general – need not be deterministic as assumed in the no-deterministic-but-indeterministic intervention case. However, one could insist that the definition of an intervention variable at least appears to presuppose that an intervention variable $I$ for $X$ is a deterministic cause of $X$, because the formulation that '$I$ acts *as a switch* for all the other variables that cause $X$' (cf. Woodward 2003: 98, emphasis added), as it stands, seems to refer to deterministic causation only. If that were a correct way of understanding Woodward's definition, then the uranium case would be more problematic for interventionists. But be that as it may. A charitable reading of Woodward's definition of an intervention should allow for indeterministic interventions.

At any rate, the no-deterministic-but-indeterministic intervention case is only problematic conditional on the assumption that only deterministic physically possible interventions are allowed. In contrast, the neither-deterministic-nor-indeterministic intervention case is unconditionally trouble for Woodward (as Woodward knows well), because there is absolutely no (neither deterministic nor indeterministic) physically possible way to intervene. For this reason, I will concentrate on the neither-deterministic-nor-indeterministic intervention case.

What do these counter-examples show? One might be inclined to object: 'Isn't the Big Bang case a rather far-fetched counter-example from physics?' However, even if these counter-examples are far-fetched and physics-based, they should raise the following suspicion: in general, and apart from these specific cases of the counter-examples, the possibility to intervene on the cause $X$ seems to be completely irrelevant for the truth conditions of '$X$ causes $Y$'. Apart from this concern over the specific choice of counter-examples from physics, the force of the arguments to be presented does not necessarily depend on the existence of actual cases of the general form of the counter-example. All concerns should still be justified even if it turned out that, as Woodward acknowledges, it is merely a 'logically coherent possibility' (Woodward 2003: 130) that $X$ is a cause of $Y$ but there is no physically possible intervention $I = i$ on $X$ – even if there are no actual (physical) cases of this kind.

Woodward rejects these kinds of counter-examples by pointing out that interventions need not even be physically possible. Woodward (2003: 128f.) understands the relevant sense physical possibility as follows:

An intervention $I = i$ is physically possible iff $I = i$ is consistent with some set of possible initial conditions and the actual laws.[5]

Woodward (2003: 128f.) claims that the counter-examples show that an intervention on $X$ is not be required to be physically possible. Instead of being physically possible, interventions are required to be possible in the sense that they are merely 'logically possible' or not 'ill-defined for conceptual or metaphysical reasons' (Woodward 2003: 128, 132). Note that, although Woodward introduces a distinction of logical, conceptual, and metaphysical possibility, he refers only to logical possibility in the large majority of cases and, further, he seems to use these kinds of modality interchangeably. For this reason, the potential differences between these kinds of modality will not be explored here.

Is Woodward's strategy to deal with the counter-examples by weakening the required notion of possibility really convincing? The next section will present two arguments against Woodward's strategy to deal with the counter-examples.

## Two arguments against the need for logically possible interventions

The first argument against Woodward's strategy aims to establish that interventions are dispensable for Woodward's semantic project provided one allows for them to be merely logically possible. The second argument supports the claim that interventions with this modal character lead to the fatal result that interventionist counterfactuals are evaluated inadequately.

### Argument 1

The first argument addresses the claim that merely logically possible interventions can be dispensed with – which would lead to a collapse of the interventionist theory as it stands. When I claim that merely logically possible interventions are dispensable, I mean that interventions with this modal character fail to contribute non-trivially to the truth conditions of causal claims; that is, interventions can be eliminated without loss. The conclusion of the first argument is that interventions are dispensable for stating the truth conditions of causal claims such as '$X$ is a direct type-level cause of $Y$'. In other words, the first argument is a dispensability argument. I use the notion of dispensability in analogy to its use in indispensability arguments in philosophy of mathematics (cf. Putnam 1975; Field 1980): a term in a theory (in our case, the notion of an intervention in a theory of causation) is dispensable if this term can be eliminated without compromising the strength of the theory (that is, the adequacy of an explication of causation).

Suppose that the causal claim 'the Big Bang is a cause of $Y$' is true (for some $Y$). According to the interventionist theory, 'the Big Bang is a cause of $Y$' is true iff the following two interventionist counterfactuals are true: (a) 'if there were an intervention $I = i$ on the Big Bang such that the Big Bang occurred as it actually did, then $Y = y$ would be the case as it actually is', and (b) 'if there were an intervention $I = i^*$ on the Big Bang such that the Big Bang were to occur differently than it actually did, then $Y$ would take a counterfactual value $y^{*}$'. However, the existence of a logically possible intervention seems to be completely dispensable for stating the truth conditions of 'the Big Bang is a cause of $Y$'. Let us begin to argue for this claim by way of an example of an intervention on the Big Bang suggested by Tim Maudlin. Having pointed out that it is neither practically nor physically possible to intervene on the Big Bang, Maudlin presents an instructive example of a physically impossible[6] intervention on the Big Bang:

> It is surely *physically* impossible that the Big Bang have [sic] been controlled [...] but it nonetheless has *effects*. Physical considerations [...] may leave the philosophical mind unimpressed. It is at least *metaphysically* possible to control the Big Bang. God might decide, for example, to employ the following scheme: if He is forgiving the universe will be open, ever-expanding; if He is jealous the universe will be closed. (Maudlin 2002: 149f.)

Maudlin's key idea can be put as follows: although there is no physically possible intervention on the Big Bang, *God* (italics are used for variables) is a logically possible (deterministic or indeterministic) intervention variable on the variable *Big Bang* relative to the variable *Evolution of the Universe* (after the Big Bang).[7] Maudlin's scenario can be depicted by Figure 4.2.

In Figure 4.2, the variables *God*, *Big Bang* and *Evolution of the Universe* have the following possible values:

- The variable *God* ranges over the set of possible values {forgiving; jealous};
- The binary variable *Big Bang* ranges over the set of possible values {1, 2}. These values represent two different ways in which the Big Bang might have taken place, as Maudlin's scenario suggests that the two possible moods of god (i.e. to be jealous and to be forgiving) lead to different initial conditions of the universe;
- The variable *Evolution of the Universe* has the values {open, closed}.

Let us re-describe Maudlin's divine intervention scenario in terms of interventions on *Big Bang*.

1. *God* is a cause of the *Big Bang*;
2. There is at least some value of *God* such that if *God* takes this value, then *Big Bang* depends only on *God* and *Big Bang* depends on no other variables; that is, *God* is the only cause of *Big Bang*;

*Figure 4.2* Causal structure of the Big Bang scenario

3. *God* is not a direct cause of *Evolution of the Universe*, and if *God* is a cause of *Evolution of the Universe* then *God* is an indirect cause of *Evolution of the Universe* via a causal path leading through *Big Bang* and a – possibly empty – set of intermediate variables $Z_1, \ldots, Z_n$;

4. *God* is probabilistically independent of other causes $W_1, \ldots, W_n$ of the *Evolution of the Universe*, which are not on a causal path leading from the *Big Bang* to the *Evolution of the Universe*;

5. *God* does not alter the relationship between *Evolution of the Universe* and its causes $W_1, \ldots, W_n$, which are not on a causal path leading from *Big Bang* to *Evolution of the Universe*.

If *God* takes either of the values in its range, then the probability increases that the Big Bang takes place in a different manner (as indicated by *Big Bang*'s range {1, 2}) and different evolutions of the universe result (as indicated by *Evolution of the Universe*'s range {open, closed}). All of this is in perfect accordance with Maudlin's story about the possible scenarios. Thus, each of the statements *God = forgiving* and *God = jealous* represents a merely logically possible intervention on *Big Bang* relative to *Evolution of the Universe*. Maudlin's scenario illustrates a merely logically possible intervention on the Big Bang.

Clearly, interventionists are committed to accept the presented intervention – because it is logically possible that there is something about the moods of God that has the desired intervening effect on the Big Bang. Obviously, it is easy to come up with many other examples of merely logically possible interventions (e.g., the physical constants case and Norton's restricted dome). If this is correct, then a problem arises for the interventionist: once one admits merely logically possible interventions, interventions become semantically dispensable. Let us argue for this claim by focusing on the Big Bang scenario: suppose that the causal claim '*Big Bang* is a cause of *Evolution of the Universe*' is true. Interventionists claim that '*Big Bang* is a cause of *Evolution of the Universe*' is true, roughly, iff the following two interventionist counterfactuals are true: (a) 'if there were an intervention *God = forgiving* on *Big Bang* such that the *Big Bang = 1*, then it would be the case that *Evolution of the Universe = open*', and (b) 'if there were an intervention *God = jealous* on *Big Bang* such that the *Big Bang = 2*, then it would be the case that *Evolution of the Universe = closed*'. In this scenario, the existence of a logically possible intervention apparently becomes dispensable for stating the truth conditions of '*Big Bang* is a cause of *Evolution of the Universe*', because the existential claim 'there is a logically possible intervention on *Big Bang*' (say, in the second counterfactual) does not add anything substantial to the following counterfactual (in which the existential claim about the intervention is eliminated):

> if it were the case that *Big Bang = 2*, and other causes of *Evolution of the Universe*, $W_1$, ..., $W_n$, had been held fixed (if there are any), then it would be the case that *Evolution of the Universe = closed, and* it is not a logical contradiction to say that the *Big Bang = 2*.

To say the least, it is not obvious what one gains by adding an existential claim about a merely logically possible intervention to the conditional above. We seem to be able to provide a complete description of the causal information contained in the Big Bang scenario by simply using counterfactuals that do not refer to interventions. However, the argument for the dispensability of Woodwardian interventions is ultimately justified in three ways. Let me provide a sketch of the dispensability arguments that will occupy me for the rest of the book.

First, if we can find a successful semantics for counterfactuals that does not rely on Woodwardian interventions, then Woodwardian interventions are dispensable for stating the truth-conditions of counterfactuals. Several standard accounts of semantics for counterfactuals are, in fact, able successfully to evaluate the conditional above without employing the Woodwardian notion of an intervention (these standard accounts are presented in the Argument 2; recent improvements of the standard accounts include Pearl 2000; Schaffer 2004b; Leitgeb 2012).[8] Thus, Woodwardian interventions are dispensable for stating the truth-conditions of counterfactuals.

Second, if a counterfactual theory of causation – that makes use of these (improved) standard semantics for counterfactuals – is able to satisfy the criteria of adequacy for a theory of causation at least as well as the interventionist theory, then interventions seem to be dispensable for an adequate theory of causation. There are such theories of causation. Thus, interventions are dispensable.

Let me provide some details and examples (see Chapter 1). For instance, an adequate theory of causation is expected to satisfy the following criteria of adequacy:

- An adequate theory ought to explicate several kinds of causation that are referred to in the sciences (e.g. actual causes, type-level causes, deterministic and indeterministic causes, contributing and total causes).
- An adequate theory of causation has to account for several typical features of causation (e.g. the time-asymmetry of causation,

the sensitivity to changes in background conditions, a distinction between genuine causation and accidental correlation).

- It is a criterion of adequacy to provide a successful description of intuitively possible causal scenarios (such as pre-emption scenarios and common-cause scenarios).

In Chapter 8, an argument will be presented for a counterfactual theory of causation – the comparative variability theory – which preserves crucial features of the interventionist account (without relying on interventions, of course). Most importantly, my counterfactual approach adopts a key feature of Woodward's theory that guarantees its (limited) success: specific background conditions (i.e. variables that are not on a directed path from the cause variable to the effect variable) have to be held fixed in a possible situation in which the cause variable takes a counterfactual value.[9] The upshot of my approach is that causal dependence amounts to conditional counterfactual dependence – that is, conditional on a certain fixed causal background, a causal field (cf. Field 2003: 452). However, this feature of holding fixed the causal background can be preserved even if the notion of an intervention is eliminated. I argue that the comparative variability theory is able to satisfy the criteria of adequacy listed above at least as well as the interventionist theory (and, in some cases, even better). Thus, interventions are dispensable. However, this result does not exclusively depend on my particular theory of causation. Other counterfactual theories that do not employ the strong Woodwardian notion of an intervention (such as Field 2003; Schaffer 2004b; Halpern and Pearl 2005; Menzies 2007) can also be used to establish the dispensability argument.

Third, Argument 2 will argue that there is no easy way out for interventionists: it is unfortunately not the case that interventions are dispensable but harmless. Woodwardian interventions are not harmless because they lead to a severe problem regarding the evaluation of counterfactuals.[10] If interventions create their own problems (given that we can state the truth-conditions of counterfactual conditionals and causal claims without referring to interventions), then we have an additional reason to eliminate interventions without regret.

I conclude from these three considerations that intervention can be eliminated without loss. In other words, they are dispensable. Tim Maudlin suggests a similar conclusion with respect to his scenario of a physically impossible divine intervention on the Big Bang:

The temptation to use these fantastic theological scenarios does not arise from any deep conceptual connection between causation and signals [or 'controllability' as Maudlin also says – these expressions are equivalent with possible interventions in our case]. *The scenarios rather illustrate a counterfactual connection: had the Big Bang been different so would the later course of the universe have been.* (Maudlin 2002: 150, emphasis added)

The upshot of discussing Maudlin's divine intervention scenario is that merely logically possible interventions turn out to be eliminable without loss. As Maudlin points out, the decisive issue is counterfactual dependence, which can be stated without reference to (merely logically possible) interventions.

### Argument 2

The second argument against logically possible interventions relies on three alternative approaches to the semantics of counterfactual conditionals. Let me briefly motivate why determining the truth conditions of counterfactuals matters for our present concerns. Let us recall the motivation for employing interventions. The main motivation for introducing interventions is closely tied to counterfactuals. Woodward's definitions of causation imply a specific kind of counterfactuals whose arguments are propositions that some variable takes a certain value in its range: 'if there were an intervention $I = i$ on $X$ such that $X = x$, then $Y = y$ would be the case' (cf. Woodward 2003: 15). In Argument 1, it was stated that logically possible interventions are dispensable for the truth conditions of causal statements. In Argument 2, I will change the perspective: for the sake of argument, I will grant that merely logically possible interventions do play an indispensable role for the interventionist theory. I argue that Woodward has a price to pay for this claim: merely logically possible interventions lead to the fatal result that interventionist counterfactuals involving existential claims about merely logically possible interventions are inadequately evaluated as false. If this reasoning were sound, this would also be an unwelcome result for interventionists. Let this be known as the *problem of inadequate evaluation*. It will be shown that the problem of inadequate evaluation is fatal for the interventionist theory of causation in that the problem of inadequate evaluation remains a problem for these standard semantics. It is further argued that, even if one relies on one of a variant of interventionist semantics, one fails to provide a satisfying solution to the problem of inadequate evaluation, because relying on interventionist semantics is ad hoc, or non-practical, or subject to the dispensability argument. For brevity's sake, let us restrict

Argument 2 to the strong case of merely logically possible interventions on the Big Bang.

The modal character of logically possible interventions is not as innocent as Woodward suggests, because possible worlds in which interventions of this kind exist deviate strongly from the actual world with respect to the laws of nature. This is problematic for interventionists because: (a) interventionists crucially rely on (interventionist) counterfactuals, and (b) worlds with strongly different laws are ruled out for the evaluation of counterfactuals by the three standard approaches to the semantics of counterfactuals. Consequently, if we evaluate counterfactuals in worlds where merely logically possible interventions on the antecedent variable exist, then the counterfactuals are false (problem of inadequate evaluation).

Let us explore this claim in greater detail. Assume there is a world $w$ such that a merely logically possible intervention on the Big Bang exists in $w$. By definition, worlds where merely logically possible interventions exist are physically impossible worlds; that is, worlds differing in laws from the actual world. According to our example, a world where an intervention is carried out on the Big Bang instantiates laws of nature that differ from the actual laws (namely, it is in accordance with these other-worldly laws that there is a deterministic or an indeterministic intervention on the Big Bang). Yet, is it the case that any violation of actual laws in a world $w$ rules out $w$ for the evaluation of counterfactuals (which are true or false in the actual world)? The answer is 'No'. One might wonder whether a logically possible intervention is just a local violation of a law. For this reason, an intervention is sometimes (cf. Woodward 2003: 135f.) compared with a small miracle in the similarity heuristic of Lewis's semantics (cf. Lewis 1979: 47f.). According to Lewis, the closest possible worlds are those that differ minimally from the actual world in particular matters of fact such that a counter-nomological (i.e. violating the laws of the actual world) event $m$ occurs just before the antecedent event $c$ (the cause) occurs. The event $m$ is a small miracle because after the occurrence of $m$ (i.e. after setting new initial conditions) the world again conforms to the actual laws. But note that this comparison is not wholly justified: although interventions and small miracles are both changes of particular matters of fact in possible worlds, Lewis does not argue that small miracles are causes of the antecedent event $c$ in question. And it is precisely the causal character of interventions – which interventionists explicitly highlight – that leads to trouble.

There is a further and even more important reason to believe that an intervention on the Big Bang has to be clearly distinguished from a small miracle. In the closest worlds in which the actual laws are preserved to a maximal degree, there is no small miracle (i.e. no set of counterfactual initial conditions) such that a state of the world evolves in which the Big Bang occurs differently than it actually does. In other words, such an intervention is not even possible in the sense of physical possibility. As opposed to that, worlds in which merely logically possible interventions exist instantiate physically impossible (viewed from the actual world) nomic connections between the intervention variable *God* and the cause variable *Big Bang*. Woodward seems to be committed to these physically impossible nomic connections: he argues that true (actual and type-level) causal statements are 'backed up' by at least non-strict, invariant generalizations (cf. Woodward 2003: 146f., 244; also his definition M of causation, p. 59, implies invariant relation between causal relata; pp. 245–55 on the claim that causal relations are backed up by invariant generalizations). In other words, there is an invariant (or, in some cases, nomic), repeatable connection between a type of intervention (e.g. *God* = *forgiving*) and a type of cause (e.g. *Big Bang* = 1) in world *w*. Thus, if it is true in a world *w* that there is an intervention on *Big Bang* (i.e., according to the definition of an intervention, at least '*God* is a cause of *Big Bang*' is true in *w*), then there is an invariant generalization (or law) in world *w* that 'backs up' this causal relation between the intervention variable *God* and the cause variable *Big Bang*. The conclusion can be drawn that interventions differ from small miracles. Merely-logically possible-intervention-worlds are worlds with truly different, non-actual laws in the sense that they are counter-legal to a higher degree than are small-miracle-worlds.

This result is quite problematic: according to the three standard semantics of counterfactuals, worlds truly differing in laws from the actual world are not adequate to evaluate counterfactuals.

1. According to Lewis's possible worlds semantics, a counterfactual is (non-vacuously) true iff the consequent is true in the closest antecedent-worlds (cf. Lewis 1973a: 16). Worlds that differ in laws to a higher degree than small miracle worlds are not among the closest antecedent-worlds that are relevant for fixing the truth conditions of counterfactuals (cf. Lewis 1979: 47f.). Thus, (merely logically possible) intervention-worlds are not among the closest worlds; that is, they are big-miracle worlds involving global violations of actual laws.

2. According to the meta-linguistic account (Goodman 1983), a counterfactual is true iff the consequent can be logically derived from a set

of premises consisting of (a) laws of nature, and (b) the antecedent and other initial conditions. Using truly non-actual law statements as premises seems to be misguided, if one would like to evaluate a counterfactual in the actual world. According to the Goodmanian approach, it is the case that if a counterfactual is supposed to be true in the actual world, then the consequent has to be inferred from (maximally) actual laws (that do not include counterfactual invariant connections between a merely logically possible type of intervention and a type of cause) and singular statements.

3. According to the suppositional theory (Ramsey 1929: 247; Adams 1975, ch. 4; Skyrms 1994; Edgington 2008), a conditional is acceptable (to the degree that) if one 'hypothetically' adds the antecedent to one's knowledge, then one also believes that the consequent is true (the Ramsey test).[11] Beliefs about which laws are true in the actual world are important for carrying out the Ramsey test (cf. Leitgeb 2012, section 2). Supposing that $p$ and adding $p$ to a stock of beliefs including truly non-actual laws does not make sense. The relevant epistemically possible worlds (in a communicative situation) are at least partly constituted by maximally actual laws (that, again, do not include counterfactual invariant connections between a type of a merely logically possible intervention and a type of cause).

To sum up, merely logically possible worlds in which there are interventions on the Big Bang are truly counter-legal (i.e. they are counter-legal to a higher and more global degree than small-miracle worlds). This result leads to a conflict with the three standard semantics of counterfactuals: if we evaluate a counterfactual in worlds with truly non-actual laws, then we have to evaluate this counterfactual as false. Therefore, worlds where interventions on the Big Bang exist are not appropriate for evaluating counterfactuals: they lead to inadequate results (problem of inadequate evaluation). For instance, we have to say, from an interventionist point of view, that the counterfactual 'if the Big Bang had occurred differently, then the universe would have evolved differently' is false, because interventionists reformulate this conditional by adding an existential claim about a merely logically possible intervention on the Big Bang to the antecedent of the conditional.

Now, the burden of proof is on the interventionists' side: interventionists have to show (a) why the standard approaches are wrong in rejecting merely logically possible interventions, and (b) how to evaluate counterfactuals with a semantics that includes interventions. This is, of course, an unpleasant result for interventionists.

Yet, let us take a step back. Maybe the situation is not so unpleasant for interventionists after all. A natural response to Argument 2 could consist in adopting an interventionist semantics of counterfactuals. As mentioned, an interventionist might propose an interventionist version of possible worlds semantics along the following lines: a counterfactual is true iff the consequent is true in the closest antecendent-worlds. The closest worlds are those worlds in which the antecedent is true by virtue of an intervention on the antecedent-variable (see Chapter 3 for a detailed discussion of this semantics, as well as for Goodmanian and suppositionalist versions of interventionist semantics).

Although turning to interventionist semantics might be a tempting option for an advocate of an interventionist theory of causation, this strategy does not help to refute Argument 2 for three reasons.

First, adopting an interventionist semantics comes close to an ad hoc move in order to vindicate the interventionist theory of causation. It might be correct that the interventionist semantics can handle counter-examples to Lewis's original semantics of counterfactuals. Dealing with these counter-examples successfully is a prima facie reason in favour of interventionist semantics. However, the interventionist semantics does not explicitly presuppose a specific modal character of interventions. Building the problematic assumption of the existence of merely logically possible interventions into the semantics of counterfactuals seems to be motivated by one goal only: one excludes cases in which intervening is physically impossible (such as the Big Bang scenario) as counter-examples to the interventionist theory. Therefore, legitimizing merely logically possible interventions by incorporating them into the interventionist semantics is an ad hoc move to defend the interventionist theory, and therefore it is not convincing.

Second, Woodward argues that his interventionist theory is a 'practical' theory of causation – by contrast Lewis's counterfactual theory and Dowe's conserved quantity theory are 'impractical' theories (cf. Woodward 2003: 28–38). According to Woodward, a theory of causation is 'practical' if it succeeds in 'connect[ing] causal knowledge with some goal that has practical utility' (cf. Woodward 2003: 30). Examples of a goal that has practical utility are successful manipulation (for instance, in medical, political, and scientific contexts) and reliable experimentation in the sciences. So, Woodward claims that if we come to believe that '$X$ causes $Y$' and we understand the truth conditions of this claim in accordance with interventionist theory, then we can easily connect this causal knowledge to practical goals such as manipulation and experimentation. For instance, if we endorse an interventionist theory of causation, we know that we are, in principle, able to manipulate the effect $Y$ by manipulating the cause $X$. We also know that if we want to design an experimental set-up that involves $Y$, then the belief that $X$ causes $Y$ is indeed relevant (for instance, because we might want to isolate $Y$ from its cause $X$, or because we might want to exploit the causal relation between $X$ and $Y$ for manipulation within the experiment). However, even if one agrees with Woodward that theories of causation ought to be practical in the sense he suggests, a problem arises for the interventionist: if one allows interventions to be merely logically possible, then the interventionist theory loses its practical character. For example, suppose that it is merely logically possible to intervene on the Big Bang, and suppose that the statement 'the Big Bang causes event $e$' is true: it is hard to see which practical goal can be achieved by this causal knowledge. For instance, if we had the goal of manipulating $e$, then the interventionist theory does not explain how we can realize this goal because it is merely logically possible to change $e$ by intervening on the Big Bang. I doubt that the knowledge that the occurrence of $e$ counterfactually depends on the change of the Big Bang which is the outcome of a merely logically possible intervention satisfies Woodward's own requirement that a theory of causation ought to explain why we pursue certain practical goals. Quite the opposite is the case: accepting merely logically possible interventions undermines the appraised practical character of the interventionist theory of causation, because the link to practical goals (such as successful manipulation and reliable experimental practice in the sciences) is disrupted in cases involving interventions with this modal character. Therefore, relying on an interventionist semantics that allows merely logically possible interventions is not convincing.

Third, if an interventionist attempts to refute Argument 2 by adopting an interventionist semantics, then merely logically possible interventions still remain subject to the dispensability argument. As argued, merely logically possible interventions are dispensable. If interventions are dispensable, then relying on an interventionist semantics obviously fails to reject Argument 2.

## A reconciling remark

One may still wonder whether Woodward's interventions could be of any use for determining truth conditions of causal claims. Yes, they could. But I think that this use is merely heuristic. Let me illustrate this idea, first, by briefly reconstructing Woodward's argument against agency theories of causation, and second, by turning Woodward's argument against agency theories against his own account. Woodward (2003: 123-7) accuses agency theories of causation (e.g., von Wright 1971; Menzies and Price 1993) of making the following mistake: advocates of agency theories confuse heuristic matters with the task of stating the truth conditions of causal claims. In the case of agency theories, it may be of heuristic value to use the perspective of an agent to imagine how a cause comes into existence (i.e., we bring it about by acting properly). However, although this might be a story of auxiliary value, it does not imply that agents need to play a role in the definition of causal notions and an adequate account of truth conditions of causal claims. Analogously, and also ironically, the same objection applies to Woodward: it might be the case that interventions are of heuristic use in order to imagine how and why the cause variable takes a certain value but, although this might be a helpful story, it does not imply that interventions need to play a role in the definition of causal notions and an adequate account of truth conditions of causal claims.

## Conclusion

Interventionists require that interventions be merely logically possible. This chapter has argued against this requirement. In particular, two claims were presented: first, merely logically possible interventions are dispensable for stating the truth conditions of causal claims. If this is true, then the interventionist theory, as it stands, collapses. Second, counterfactuals involving merely logically possible interventions lead to the result that interventionist counterfactuals are inadequately false. This is likewise a poor result for the interventionist theory of causation.

The chapter introduced Woodward's definition of an intervention, and then presented a counter-example to the claim that interventions have to be physically possible for analysing causal notions: there is a case where causation takes place but no intervention is physically possible. In order to reject the counter-example, Woodward claims that the possibility of interventions needs to be understood in a weaker, logical sense. The chapter argued that this strategy is highly problematic. Two arguments were presented against logically possible interventions. Argument against this requirement was made by establishing two claims: first, merely logically possible interventions are dispensable for stating the truth conditions of causal claims. If this is true, then the interventionist theory, as it stands, collapses. Second, counterfactuals involving merely logically possible interventions lead to the fatal result that interventionist counterfactuals are inadequately false. This is likewise an unwelcome result for interventionists. I conclude that if we attempt to master the tasks of explicating causal concepts and stating the truth conditions of causal claims, we best get rid of Woodwardian interventions.

# 5
# Non-Universal Laws

## Why we need a theory of non-universal laws

Chapter 4 presented arguments against the interventionist assumption that interventions are required to be merely logically possible. That is, the notion of an intervention is troubled, because *either* interventions are dispensable (with respect to stating the truth conditions for causal statements and counterfactuals), *or* the modal character leads to the incorrect semantic evaluation of interventionist counterfactuals. These conclusions create a follow-up problem for interventionists concerning laws of nature. If one accepts the naturalist criteria of adequacy (see Chapter 1, pp. 15–21), then any theory of causation for the special sciences has to account for the no-universal-laws requirement. Interventionists respect and, prima facie, satisfy this requirement as criterion of adequacy: according to the interventionist view, causation requires invariant relationships between cause and effect. These invariant relations are described by invariant generalizations, which are not required to be universal. Hence, the no-universal-laws requirement appears to be satisfied. However, the invariance of generalizations is defined in terms of possible interventions; that is, a generalization is invariant under possible interventions. This is unfortunate for interventionists, because the notion of a possible intervention is a troubled one, as argued in Chapter 4. In what follows, it will be argued for the claim that one can account for the no-universal-laws-requirement without appealing to the notion of an intervention. In turn, a defence will be offered for an alternative explication of non-universal laws in the special sciences (or 'lawish generalizations'). This alternative explication of non-universal laws will be an important building block of my comparative variability theory of causation (Chapter 8).

For the sake of clarity and consistency, I will repeat some of the main issues that have already made their appearance in Chapter 1, where we discussed the no-universal-laws-requirement, and outlined the nomothetic dilemma for causation and explanation.

Many philosophers[1] are convinced that (fundamental) physics states universal laws, while the special sciences (e.g., biology, psychology, sociology, economics, medical science and so on) state non-universal or ceteris paribus laws (henceforth, cp-laws).[2] We may consider Barry Loewer's view as paradigmatic. Loewer describes the differences between fundamental physical laws (he uses Newton's laws of motion as an example) and special science laws as follows:

> The main relevant differences between fundamental dynamical laws and special science laws are these: The candidates for fundamental dynamical laws are (i) *global*, (ii) *temporally symmetric*, (iii) *exceptionless*, (iv) *fundamental (not further implemented)*, and (v) *make no reference to causation*. In contrast, typical special science laws are (i\*) *local*, (ii\*) *temporally asymmetric*, (iii\*) *multiply realized and implemented*, (iv\*) *ceteris paribus*, and (v\*) *often specify causal relations and mechanisms*. (Loewer 2008: 154, original emphasis)

In this chapter, I will agree with Loewer and others that the dynamical laws in fundamental physics and the laws in the special sciences differ in the way they describe.[3] Nonetheless, only some of the features of special science laws will be explicitly adddressed – such as being local, having exceptions or being ceteris paribus. I will leave aside features such as the possible multiple realization of special science laws, being temporally asymmetric, and the question of whether and how fundamental physical laws relate differently to causation than special science laws (although the last two features will be discussed in Chapter 6). Despite these differences between laws in fundamental physics and laws in the special sciences, most philosophers believe that in both physics and the special sciences, laws are important because they are statements used to explain and to predict phenomena, they provide knowledge on how successfully to manipulate the systems they describe, and they support counterfactuals and so on. Statements that play these roles in the sciences I term lawish. Similarly, Mitchell (1997, 2000) characterizes generalizations in the special sciences as 'pragmatic laws' by virtue of their performing at least one these roles. Note that, in the debate on laws of nature, lawlikeness is commonly associated with universality (Braithwaite

1959: 301). In contrast, I use 'lawish' differently: a general statement is lawish if it is of explanatory and predictive use, successfully guides manipulation, and supports counterfactuals. Thereby, contrary to the traditional understanding of laws, being lawish neither requires universality nor other characteristic features of fundamental physical laws. It is certainly a matter of convention whether one would still want to use the term 'law' for non-universal general statements. One can either use a new term for lawish, non-universal explanatory, general statements (e.g., Woodward and Hitchcock 2003 introduce the term 'explanatory generalization'). Or, as I maintain in this chapter, one can insist that if a statement plays a lawish role then it shares a sufficient number of properties with universal laws to justify being called a law. Christopher Hitchcock and James Woodward admit that their account may be read as a reconceptualization of lawhood (cf. Woodward and Hitchcock 2003: 3).

Let me provide some examples of special science laws. Examples from economics are the *law of supply* and *the law of demand*, which – in the words of John Roberts, a critic of cp-laws – state (cf. Roberts 2004: 159, also Kincaid 2004: 177):

> *Law of supply*: If the supply of a commodity increases (decreases) while the demand for it stays the same, then the price decreases (increases).
>
> *Law of demand*: If the demand for a commodity increases (decreases) while the supply remains the same, then the price increases (decreases).

Another example of a lawish statement in the special sciences is the area law in island bio-geography:

> the equilibrium number S of a species of a given taxonomic group on an island (as far as creatures are concerned) increases [polynomially][4] with the islands area [A]: $S = cA^z$. The (positive-valued) constants $c$ and $z$ are specific to the taxonomic group and island group. (Lange 2000: 235f; cf. Lange 2002: 416f.)

Although there is insufficient space to discuss them here, other vividly and controversially debated examples of special science laws are: in neuroscience, the Hodgkin–Huxley model of action potential (cf. Weber 2008: 997–1001) and the generalizations describing the mechanism of long-term potentiation (cf. Craver 2007: 65–72, 168); in psychology, generalizations describing learning and memory (cf. Gadenne 2004: 107f); in economics, generalizations in models of economic growth (cf. Kincaid 2009: 456f); and in biology, Mendel's law of segregation, the Hardy–Weinberg law, and the principles of natural selection (cf. Rosenberg and McShea 2008: 36; also Beatty 1995; Sober 1997; Mitchell 1997, 2000; Rosenberg 2001).

Generalizations like these are believed to be lawish, although they are not universal generalizations. As noted, traditionally, the most important feature of a law to understand its lawlikeness is universality (cf. Lewis 1973a: 73–6; Armstrong 1983: 88–93). Furthermore, picturing lawlikeness mainly in terms of universality has led many theories of causation and explanation to rely on universal laws. It transpires that this assumption is problematic: the major challenge for any theory of non-universal laws in the special sciences is to account for their apparent lawish function (in the sense introduced above). If we are not able to provide an explication of non-universal laws, then (at least) the philosophy of the special sciences faces a severe problem concerning causation and explanation in its domains. As argued in Chapter 1 (p. 17, introducing the requirement of non-universal laws), many theories of causation (such as regularity views, counterfactual and conserved quantity theories) and explanation (such as various models of causal and mechanistic explanation) in their standard form presuppose universal laws of nature. As also outlined in Chapter 1 (pp. 19–20), the ultimate problem arising from many theories of causation and explanation can be represented by the following cluster of jointly inconsistent claims:

- The special sciences (a) refer to causes in their domains (i.e. some causal statements in these sciences are true), and (b) provide explanations in their domains.
- It is a plain fact that the special sciences – possibly, in contrast to physics – cannot rely on universal laws.[5]
- Most philosophical theories of causation and explanation – in their standard form – presuppose universal laws.

We can state the inconsistency of these three claims in the form of the nomothetic dilemma of causality and explanation (cf. Pietroski and Rey 1995, 85; Woodward and Hitchcock 2003: 2):

*First horn*: If it is a plain fact that the special sciences cannot rely on universal laws (Assumption 2), and if most philosophical theories of causation and explanation involve universal laws and we do not reject these theories (Assumption 3), then there is neither causation nor explanation in the special sciences (negation of Assumption 1).

*Second horn*: If there is causation and explanation in the special sciences (Assumption 1), and if it is a plain fact that the special sciences cannot rely on universal laws (Assumption 2), then there is causation and explanation that does not involve universal laws (negation of Assumption 3); that is, we have to reject the above-listed theories of causation and explanation in their standard form.

If we do not want to give up the immensely plausible opinion that the special sciences refer to causes and provide explanations (Assumption 1) for purely philosophical reasons, then we are in need of a theory of non-universal laws. So, let us opt for the second horn of the nomothetic dilemma.

The next section will introduce Lange's Dilemma stating that non-universal laws are either false or trivially true. Lange's Dilemma is a challenge because we expect non-universal laws to be true empirical statements. The subsequent section refers to an attempt to avoid Lange's Dilemma proposed by Paul Pietroski and Georges Rey (1995). Pietroski and Rey attempt to save a cp-law L from being trivially true by explaining counter-instances to L. Although this account has been criticized on good grounds, the basic idea is fairly correct. Even its critics, John Earman and John Roberts (1999) among others, sketch a way to repair the account by Pietroski and Rey. Earman and Roberts require that a story has to be told about the relevance of the antecedent for the consequent of a law statement (which I term the 'Requirement of Relevance'). I attempt to fulfil this requirement. The chapter continues by setting up a theory of non-universal laws by distinguishing different meanings (or dimensions) of 'non-universal'. It is then argued that this approach (a) allows avoidance of Lange's Dilemma by conceiving special science laws as quasi-Newtonian, and (b) that it meets Roberts and Earman's requirement of relevance by spelling out relevance in terms of a refined notion of invariance, invariance* (i.e. this refined notion of invariance does not rely on interventions). The chapter closes by arguing that the results of the preceding sections amount to the following theory of special science laws: L is a special science law iff (a) L is a system law, (b) L is quasi-Newtonian, and (c) L is minimally invariant*. A considerable advantage of my explication of special science laws consists in the fact that, on the one hand, it avoids the troubled notion of an intervention, while on the other hand, it satisfies the no-universal-laws requirement (which is a criterion of adequacy for any explication of causation in the special sciences (see Chapter 1, pp. 17–21).

## Challenge I: falsity and triviality

A philosophical reconstruction of lawish statements in the special sciences faces a severe problem, which can be articulated in the form of Lange's Dilemma (cf. Lange 1993: 235). Here is the first horn:

*First horn – Falsity*: Strictly and literally speaking, special science laws are false because it is not the case that all Fs are Gs (if that is what the laws say).

For instance, the relationship between supply and price is not always as the law of supply says (or, as it seems to say prima facie), because a disturbing factor might occur. In other words, special science laws that instantiate perfect regularities are – mildly put – 'scarce' (Cartwright 1983: 45). Yet, if one supposes that the law is to be formalized as a universally quantified conditional sentence, then one counter-instance (due to a disturbing factor) to the universally quantified sentence means that it is false.

The second horn of Lange's Dilemma can be stated as follows:

*Second horn – Triviality*: If laws in the special sciences are cp-laws, then they are trivially true.

If we suppose that an implicit cp-clause is attached to the law, then it seems to be equivalent to 'All Fs are Gs, *if nothing interferes*'. But then the cp-law in question is in danger of lacking empirical content. It lacks empirical content because it seems to say nothing more than 'All Fs are Gs or (it is not the case that all Fs are Gs)'. Note that the second horn of Lange's Dilemma seems to depend on the exclusive reading of the ceteris paribus clause ('if nothing interferes', 'if all disturbing factors are absent'). If this were the correct theory of laws in the special sciences, then these laws would be analytically true sentences and, therefore, trivially true. Obviously, this is a poor result because laws of the special sciences should be reconstructed as (approximately) true empirical statements – not as sentences being true by virtue of the meaning of their components. Note that the second horn is a more pressing problem than the first, as I have already given up the assumption that special science laws are universal (as presupposed in the first horn) – to be precise, I reject the claim that special science laws are universal in two readings of 'universal'. In the recent debate, some philosophers take Lange's Dilemma as a reason to be pessimistic about whether there really is a convincing explication of laws in the special sciences:

> there is no persuasive analysis of the truth conditions of such laws; nor is there any persuasive account of how they are saved from

vacuity; and, most distressing of all, there is no persuasive account of how they meld with standard scientific methodology, how, for example, they can be confirmed or disconfirmed. *In sum, a royal mess.* (Earman and Roberts 1999: 470f., emphasis added)

Dealing with this dilemma is then clearly a central semantic challenge, which also has epistemological consequences, as Earman and Roberts point out: in order to be empirically testable, special science laws have to be true (contrary to the first horn) and non-trivial (contrary to the second horn). Here, attention will be restricted to the semantic challenge posed by Lange's dilemma.

## Challenge II: the requirement of relevance

Pietroski and Rey (1995: 92) claim that it is sufficient for a law 'cp,∀(x)Ax→Cx' to avoid Lange's Dilemma (i.e. to be neither necessarily false nor trivially true) if the following conditions are satisfied:

- A and C are nomological predicates.[6]
- Assessing a law statement L 'cp,∀(x)Ax→Cx' leads to a commitment which is expressed by the Explanatory-Commitment-Condition (ECC): if a counter-instance (A∧¬C) to law statement L occurs, then one is committed to explain ¬C by referring to a factor H which is independent of L. Pietroski and Rey allow two possibilities with respect to the independent explanatory force of H: (a) H alone explains ¬C, or (b) H in conjunction with L explains ¬C.
- It is the case that either (a) A∧C, or (b) A∧¬C and ECC is satisfied.

According to Pietroski and Rey (1995: 90), ¬C is explained independently of L by referring to H if (a) H is not a logical consequence of L (i.e. logical independence of H), and (b) the explanatory factor H is not an effect of ¬C (i.e. causal independence of H). The critics of this approach have argued that ECC is not sufficient for saving special science laws from Lange's Dilemma, because ECC allows that (a) A is completely irrelevant for C and (b) ¬C is still perfectly explained independently of L by a factor H (cf. Earman and Roberts 1999: 453f.; Schurz 2001: 366f.; Woodward 2002: 310). Earman and Roberts provide the following counter-example:

Unfortunately, [Pietroski and Rey's proposal] is not sufficient for the non-vacuous truth of the cp-law. To see why, let 'Fx' stand for 'x is spherical', and let 'Gy' stand for y = x and y is electrically conductive'. Now, it is highly plausible that for any body that is not electrically conductive, there is some fact about it – namely its molecular structure – that explains its non-conductivity, and that this fact also explains other facts that are logically and causally independent of its non-conductivity – e.g., some of its thermodynamic properties. Thus, clauses (ii) and (iii) [i.e. the conditions (2) and (3) of Pietroski and Rey's account above] appear to be easily satisfied. *If Pietroski and Rey's proposal were correct, then it would follow that ceteris paribus, all spherical bodies conduct electricity.* (Earman and Roberts 1999: 453, emphasis added)

Earman and Roberts comment on their counter-example to Pietroski and Rey's approach:

The general moral of this observation seems to be that it is not enough simply to require, [...] that when cp:(A→B), any case of A accompanied by ~B must be such that there is an independent explanation of ~B. This is because this requirement does not guarantee that A is in any way relevant to B, which surely must be the case if cp:(A→B) is a law of nature. *Perhaps Pietroski and Rey's proposal could be modified to remedy this defect.* (Earman and Roberts 1999: 454, emphasis added)

My theory of non-universal laws is an attempt to explain what 'relevance' means and to save (a version of) ECC at the same time. I agree with the critics that ECC cannot be sufficient. But it will be argued that Pietroski and Rey basically had the right idea. On the one hand, it is correct that something close to ECC is necessary for a theory of special science laws: it is necessary in order to deal with disturbing factors. I will rely on quasi-Newtonian laws for these purposes. On the other hand, Earman, Roberts, Schurz and Woodward are completely justified in demanding that we have to account for the relevance of the antecedent for the consequent of a law statement. I will provide such an account of relevance in terms of a refined notion of invariance.

## Four dimensions of non-universal laws

As argued at the beginning of the chapter, it is the received view that, in the special sciences, laws appear to be non-universal – or, they are said to 'have exceptions'. But what does it mean to be universal or non-universal? Surprisingly, in the recent debate on cp-laws this question is not answered in a systematic way.[7] The lack of a systematic approach is a major problem, because universality is an ambiguous concept. In accord with Andreas Hüttemann (2007: 139–41), I distinguish four meanings or dimensions of universality with respect to a law statement:

Dimension 1 – *Universality of space and time*: Laws are universal$_1$ iff they hold for all space-time regions.

Dimension 2 – *Universality of domain of application*: Laws are universal$_2$ iff they hold for all (kinds of) objects.

Dimension 3 – *Universality for external circumstances*: Laws are universal$_3$ iff they hold under all external circumstances (i.e. circumstances that are not referred to by the law statement itself).[8]

Dimension 4 – *Universality with respect to the values of variables*: Laws are universal$_4$ iff they hold for all possible values of the variables in the law statement. Universality in this sense acknowledges that laws usually are quantitative statements (and, thus, the predicates contained in these statements are to be conceived as variables ranging over a set of possible values).

Paradigm examples of fundamental physical laws (such as Newton's laws, Einstein's field equations, and the Schrödinger equation) are usually taken to be universal in all four dimensions (cf. Schurz 2002: s. 6.1; Hüttemann 2007: 139–41). I will argue that lawish generalizations in the special sciences are (a) universal in the first and second dimension, and (b) non-universal with respect to the third and the fourth dimension of universality. This diagnosis amounts to a challenge: any theory of lawish generalizations in the special sciences is obliged to explain how a lawish statement can be non-universal$_{3\&4}$ and still play a lawish role.

## Universality$_{1\&2}$: system laws

Are special science laws universal in the first and the second dimension of universality? I believe the answer is 'Yes'. I will argue for two claims: first, lawish generalizations in the special sciences hold for all space-time regions (i.e. they are universal$_1$); however, these generalizations simply lack application in some space-time regions. Second, lawish statements in the special sciences can be reconstructed in such a way as not to quantify over a restricted domain of objects (i.e. they are universal$_2$). Arguing for these claims might not seem plausible at first glance, because generalizations in the special sciences are usually interpreted as system laws. Gerhard Schurz (2002: s. 6.1) has introduced the notion of system laws: while fundamental physical laws 'are not restricted to any special kinds of systems (be it by an explicit antecedent condition or an implicit application constraint)' (Schurz 2002: 367), system laws refer to particular systems of a certain (biological, psychological, social and so on) kind *K* in a specific space-time region. Hence, the usual characterization continues, lawish statements in the special sciences typically have an in-built historical dimension that the fundamental physical laws lack, because they are restricted to a limited space-time region where the objects of a certain kind *K* exist (for instance, cf. Beatty 1995; Rosenberg 2001).[9] I will argue that Schurz is absolutely correct in characterizing lawish statements in the special sciences as being 'restricted to [...] special kinds of systems (be it by an explicit antecedent condition or an implicit application constraint)' (Schurz 2002: 367) – however, if one adopts this characterization of special science laws as system laws, then one is still not committed to denying that these law statements are universal in the first and second dimension.

Does this characterization of system laws mean that special science laws are non-universal$_1$? No. Simply because a generalization G does not have an application in some space-time region *s*, it does not mean that the law does not hold at *s*. In order to be truly non-universal$_1$, G would have to conform to Tooley's thought experiment 'Smith's Garden':

> All the fruit in Smith's garden at any time are apples. When one attempts to take an orange into the garden, it turns into an elephant. Bananas so treated become apples as they cross the boundary, while pears are resisted by a force that cannot be overcome. Cherry trees planted in the garden bear apples, or they bear nothing at all. If all these things were true, there would be a very strong case for its being a law that all the fruit in Smith's garden are apples. And this case would be in no way undermined if it were found that no other gardens, *however similar to Smith's garden in all other respects*, exhibited behaviour of the sort just described. (Tooley 1977: 686, emphasis added)

So, according to Tooley, a law L can be spatio-temporally restricted to a space-time region *s* (as the laws in Smith's garden) in the sense that L fails to be true in a situation that is perfectly similar to the situation in *s*, except for the fact that this perfectly similar situation is located in a different space-time region *s\**.

I believe that laws that are truly non-universal$_1$ would be similar to the laws that are true of various fruit in Smith's garden. But it seems to be far too strong a claim that laws in the special sciences are local in the same way as the laws in Smith's garden. Thus, a more promising option seems to be to say that (a) special science laws are universal$_1$, and (b) these laws simply lack application in some space-time regions. For instance, to say that the law of supply does not hold on Mars because there are no people buying and selling goods does not indicate that the law of supply is a local law. A better understanding seems to be that the law of supply factually has no application on Mars (or, it lacks instances on Mars).

Does this characterization of lawish statements in the special sciences as system laws mean that special science laws are non-universal$_2$? No. At first glance, special science laws, if viewed as system laws, appear to be non-universal$_2$: special science laws quantify over a restricted domain of objects of a certain kind – not over a domain of objects of all kinds. For instance, the law of supply seems to quantify over the restricted domain of commodities – not over an unrestricted domain. So, one might get the idea that the domain of a special science is a restricted domain. The law of supply can be formalized as quantifying over a restricted domain $C$ of commodities (with $c$ as an individual variable of this domain):

$$\forall(c)((\text{supply increases})c \wedge (\text{demand constant})c) \rightarrow (\text{price decreases})c^{10}$$

But is this really a convincing reconstruction of lawish statements in the special sciences? There is an alternative formalization that quantifies over a domain of all objects. This formalization interprets the kind of object (here: commodities) as a predicate and not as a restriction of the domain (with $x$ as an individual variable for the unrestricted domain):

$$\forall(x)((\text{commodity})x \wedge (\text{supply increases})x \wedge (\text{demand constant})x) \rightarrow (\text{price decreases})x$$

The second, unrestricted formalization of the law of supply is a way to save universality$_2$. Formalizing special science laws in this form, reconstructs them as laws that hold for all objects.[11]

Obviously, I do not offer a theory for the first and second dimension of universality. Such a theory would have to elucidate why generalizations that are universal$_{1\&2}$ are lawish. All I have done is to provide a reconstruction such that special science laws can be consistently understood as being universal$_{1\&2}$. This is not a trivial result, as philosophers such as Beatty (1995) and Rosenberg (2001) insist that generalizations in the biological and the social sciences should be regarded as (a) being historical in the sense of applying only to a specific space-time region (this is in contradiction with universality$_1$), and (b) as referring to a restricted domain of objects (this contradicts universality$_2$). Contrary to these philosophers, I merely wish to point out that one can maintain that lawish generalizations in the special sciences are universal with respect to the first and the second dimension of universality. It is a matter of convention to then still call these lawish statements 'system laws' in order to highlight difference to fundamental laws.

In the following sections, I will argue that general statements play a lawish role, if they can deal with disturbing factors (Dimension 3), and if they are invariant under different possible initial conditions (Dimension 4). In other words, all the work is to be done by theories of non-universality$_{3\&4}$.

## Non-universality$_3$: the method of quasi-Newtonian laws

How shall we deal with the third dimension; that is, the fact that special science laws are sensitive to external factors? Recall that the second horn of Lange's Dilemma presupposes a reconstruction of law statements that are qualified by only a cp-clause such as '*all disturbing factors are absent*'. But one is not committed to this reading. Is there an alternative reconstruction? My positive thesis is:

> Positive thesis: A special science law L is backed up by a distinct law L* which describes (comparatively or quantitatively) the influence of relevant disturbing factors.

In order to argue for my positive thesis, I will rely on the concept of a quasi-Newtonian law (I adopt the name Tim Maudlin uses, see pp. 137f.). The basic idea is that factors leading to a counter-instance of the law L are described by another law L*. My positive thesis is intuitively supported the fact that laws are typically not isolated but are part of a theory or a model. Some disturbing factor with respect to law *L* is often described by law *L** in the same theory or model. The positive thesis is a version of ECC. The claim that the influence of a disturbing factor (for a lawish statement L) is described by another law L* is fairly close to (or a special case of) explaining the counter-instances of L by an independent factor – as ECC states. Yet, ECC and quasi-Newtonian laws differ in at least one important respect: while Pietroski and Rey's ECC seems to aim at epistemic acceptability conditions of a cp-law, quasi-Newtonian laws provide non-epistemic and perfectly objective truth conditions for laws in the special sciences.

Historically, this key idea of dealing with disturbing factors has been proposed by John Stuart Mill:

> *The disturbing causes have their laws, as the causes which are thereby disturbed have theirs*; and from the laws of disturbing causes, the nature and amount of the disturbance may be predicted a priori, like the operation of the more general laws which they are said to modify or disturb, but with which they might more properly said to be concurrent. (Mill 2008 [1836]: 50, emphasis added)

For instance, the law of supply states 'if the supply of a commodity increases (decreases), then the price decreases (increases)'. It is usually added to the antecedent '... *while the demand for this commodity stays the same*', which implies that the law of supply does not hold if the demand increases or decreases. At this point, it is crucial to notice that the evolution of the price of a good is not described by a single generalization; that is, by the law of supply. The evolution of the price also depends on another factor, the demand for a good, described by the law of demand: 'if the demand for a commodity increases (decreases) *while the supply remains the same*, then the price increases (decreases)'. It has to be emphasized that the equilibrium model of supply and demand also describes what would happen, if the demand does not remain the same. In other words, the evolution of the price of a commodity is described by an equilibrium model, according to which supply and demand can vary independently (Hausman 1992; Mas-Colell et al. 1995).

In order to illustrate Mill's original idea how a disturbing factor can also be described by a law, I draw on Maudlin's (2004) concept of a quasi-Newtonian law. In Newton's physics, Newton's First Law describes the inertial behaviour of a physical system: the uniform motion of a system when no force acts upon it. Newton's Second Law describes the deviant behaviour: the change of inertial motion if other forces are present. Maudlin characterizes the general form of Quasi-Newtonian Laws by analogy to Newton's laws:

> Let us denominate laws *quasi-Newtonian* if they have this form: There are, on the one hand, *inertial laws* which describe how some entities behave when nothing acts on them, and there are *laws of deviation* that specify in what conditions, and in what ways, the behavior will deviate from the inertial behavior. (Maudlin 2004: 431, emphasis added)

Maudlin (2004: 434) stresses that special science laws are typically quasi-Newtonian. So, let us apply Maudlin's idea to the economic case: the law of demand describes inertial behaviour; if the law of supply is integrated in the equilibrium model, then the whole model describes the deviant behaviour (of the price). Thus, the laws of demand and supply describing the evolution of the price of a commodity are quasi-Newtonian laws. As Lange observes, the case is analogous concerning the area law in island bio-geography: it matters how far an island is away from the coast – being very far away might count as a disturbance of the area law. Island bio-geographers describe this disturbing factor with the distance law: '*ceteris paribus*, islands farther away from the mainland equilibrate at lower biodiversity levels' (Lange 2002: 419). In Maudlin's terminology, we might call a model including the distance law a law of deviation (with respect to the area law as an inertial law), a quasi-Newtonian law of island bio-geography. In a recent paper, Craig Callender and Jonathan Cohen draw a similar analogy between Newton's laws of motion and the way in which the special sciences (they focus on Malthus's exponential law in ecology) describe the deviation with respect to one law by other laws:

> Consider the so-called first principle of population dynamics, Malthus's exponential law: $P(t) = P_0 e^{rt}$, where $P_0$ is the initial population (say, of rabbits), $r$ the growth rate, and $t$ the time. This ecological generalization is very powerful. It supports counterfactuals and crucially enters ecological predictions and explanations. It has an undeniably central role in most presentations of the science of ecology. *Indeed, it arguably has the very same central role in ecology that Newton's first law does in classical mechanics: both express a kind of ideal default behavior – exceptions to which are to be explained with further laws* (Ginzburg and Colyvan, 2004). In short, pending some good reason for distinguishing Malthus's Law over Newton's first law, there is good reason for taking seriously the idea that the former should count as a bona fide law of nature. (If you don't like 'law of nature', substitute 'widely applicable projectible generalization' in its stead.) (Callender and Cohen 2010: 427, emphasis added)

Maudlin calls these laws quasi-Newtonian for a good reason, because there are important disanalogies to Newton's laws and special science laws (if understood in terms of quasi-Newtonian laws):

First, in the economic case (as well as in the island bio-geographical and the ecological cases) it depends on pragmatic choice whether the law of supply or the law of demand is dubbed 'inertial law'.[12] The important point for our purposes is that a larger model in which the inertial law is integrated (e.g. the equilibrium model of a market) describes that deviating behaviour of the price evolution – that is, the behaviour is deviating relative to a chosen inertial law such as the law of supply.

Second, in cases of special science laws the deviation laws are usually not universal$_{1-4}$ as Newton's Second Law is.

Third, and most importantly, those disturbing factors governed by the laws of deviation fall into two classes: those that are in the scope of a particular discipline, and those that are not. For a special science, such as economics, there will always be disturbing causes (such as comets) which will not be integrated into the models of this discipline. Concerning these latter factors, we are committed to an existential claim if we wish to maintain that special science laws are quasi-Newtonian: if a special science system deviates from its inertial behaviour, then there are (known or unknown) laws of deviation describing the influence of disturbing factors on the inertial behaviour.[13] The examples introduced in the preceding paragraph provide a good reason to believe that this existential claim is likely to be true, because the influence of disturbing factors within a specific special science (e.g. micro-economics, bio-geography, ecology) is, indeed, described by a law of deviation. Since the scope of a special science is limited (e.g. influence of comets crashing on Earth is not described within economics), an advocate of the quasi-Newtonian approach has to make the amendment that there are – unlike in the case of Newtonian laws of motion – *unknown* disturbing factors and *unknown* corresponding laws of deviation from the point of view of a special science such as economics. Economists, for instance, seem to refer to such unknown laws of deviation when they talk about 'externalities'. Michael Strevens (2010: s. 3) argues for a similar point: the conditions of application of a special science law are partly 'opaque' for the researchers, say economists, because the researchers lack complete knowledge of all disturbing factors and the laws of deviation governing them. However, I agree with Strevens that the fact that economists have incomplete knowledge does not imply that the existential claim about unknown disturbing factors and unknown corresponding laws of deviation is false.

Despite these disanalogies, I think that the positive analogy remains intact: some laws in physics and in the special sciences are quasi-Newtonian because the influence of a disturbing factor on a system described by a law L is described (comparatively or quantitatively) by another law L*.

Relying on quasi-Newtonian laws raises a question: do scientists need to know all the laws of deviation describing the influence of all possible disturbing factors? Are all (possible) disturbing factors equally important? And, if this is not the case: how does one distinguish the important disturbing factors from the irrelevant? Intuitively speaking, it seems quite obvious (and descriptively adequate) that scientists are not interested in all possible disturbing factors. Rather, scientists seem to discriminate relevant and (more or less) irrelevant disturbing factors. Marc Lange (2000) provides a convincing pragmatic answer to the questions asked above. According to Lange's core idea, laws in the special sciences are propositions whose application is pragmatically restricted to the purposes of a scientific discipline (Lange 2002: 416). Lange identifies several strategies that scientists use in order to distinguish relevant disturbing factors from irrelevant ones. For instance, Lange discussed what one might call the 'strategy of non-negligibility': instead of providing a complete list of all interfering factors, scientists merely refer to those interfering factors 'that arise sufficiently often, and can cause sufficiently great deviations from $G$-hood, that a policy of inferring $F$s to be $G$ [...] would not be good enough for the relevant purposes' (Lange 2000: 170; 2002: 411). For instance, consider the economic law of supply. According to Lange, it may happen that the increase in supply is so small that no decrease in price results. It depends on the goals of the researchers in question whether this description of the increase in supply is sufficiently fine-grained. It might also happen that the price does not decrease while the supply increases significantly, because a gigantic comet hitting the planet Earth and destroying all life on its surface disturbs the instantiation of this law. The comet causes sufficiently great deviation from a decrease in the price of a good. Nevertheless, comets are negligible for the purposes of economists because their occurrence does not arise sufficiently often to count as a disturbing factor that is to be explicitly listed in the cp-conditions of the law of supply.

The lesson we can learn from Lange is that, ontologically speaking, all of the disturbing factors (and the corresponding laws of deviation) are on a par. However, from a pragmatic point of view, it seems to be the case that scientists rank the relevance of disturbing factors with respect to the aims of the research in their particular fields. If one follows Lange, then one way to describe this ranking is, for instance, the strategy of non-negligibility. If one considers this strategy to be the sensible reconstruction of a methodology implicit in scientific practice, then only those disturbing factors that are evaluated as relevant in the light of this strategy have to be described by laws of deviation.

## Non-universality$_4$: invariance

My earlier diagnosis was that special science laws might fail to hold for all possible values of the variables in the law statement. That is, they are non-universal$_4$. According to invariance theories of laws, a generalization may be non-universal$_4$ but nonetheless lawish. The general idea of invariance theories of laws consists in the claim that only laws remain true for different possible initial conditions that are taken to be the results of interventions. A generalization is invariant if it holds for some, possibly limited, range of the possible values of variables (expressing initial conditions that are taken to be results of interventions). For instance, the law of demand might be true for various possible units of a commodity that are demanded, but it might not hold for extremely high amounts of demand (e.g., for a demand of 30 billion units of any commodity; this feature is often referred to as the elasticity of the price). The decisive question is whether special science laws can be lawish and still be non-universal$_4$? I think that invariance theories of laws provide a positive answer to this question. Furthermore, I argue that invariance theories have an additional benefit: they satisfy the requirement of relevance, because relevance can be spelled out as invariance. I will start by using Woodward and Hitchcock's invariance theory of laws.[14] Then, I will propose a revised version of an invariance theory, because Woodward and Hitchcock's theory crucially depends on the notion of possible intervention, which is regarded as problematic.

According to Woodward and Hitchcock (2003: 17) and Woodward (2003: 250), a statement G is minimally invariant iff the testing intervention condition holds for G. The testing intervention condition states for a generalization G of the form $Y = f(X)$:

1. there are at least two different possible values of an endogenous variable $X$, $x_1$ and $x_2$, for which $Y$ realizes a value in the way that the function $f$ in G describes; and
2. the fact that $X$ takes $x_1$ or, alternatively, $x_2$ is the result of an intervention.

Let me first turn to part 1 of the testing intervention condition. The most intuitive case of a testing intervention might be the following: $X = x_1$ describes an actual state of affairs while $X = x_2$ describes a possible counterfactual state of affairs. For instance, suppose that the ideal gas law – pV = NkT – is true for the actual temperature of a gas $g$ of 30°. According to the first part of the testing intervention condition, the ideal gas law is minimally invariant if it also holds for a (counterfactual) temperature of, say, 40°.

Analogously, the area law and the law of supply conform to the first part if these laws remain true for at least two values of the independent variables (i.e. the island area variable and the supply variable).

Part 2 of the testing intervention condition specifies and restricts the type of situation in which part 1 is supposed to hold: it is a (counterfactual) situation in which the behaviour described by the generalization in question is undisturbed by external factors. For instance, in the case of the law of supply one considers a situation in which the demand stays constant (i.e. the demanded quantity of a commodity is in the 'redundancy range' – the demand is held constant such that the price is not influenced by the demand, if the supply does not change, cf. Hitchcock 2001: 290; Woodward 2003: 83). Given such a situation, one changes the values of a variable as indicated by part 1 of the testing intervention condition. Woodward and Hitchcock prefer to describe the change of a variable causally: an intervention causes the change of a variable in a (counterfactual) situation in which the behaviour described by the generalization in question is undisturbed by external factors in the sense that disturbing causes are held constant in the redundancy range (for a precise definition of intervention variables and interventions: cf. Woodward 2003: 98; and Chapter 2, pp. 29–32).

Yet, I think that the reference to interventions is by no means necessary in order to state what invariance is (cf. Mitchell 2002: 346f., for a similar view). And, as the conclusion of Chapter 4 suggests, a theory of special science laws ought not to use the troubled notions of a Woodwardian intervention. An alternative way to characterize invariance would be analogously to the testing intervention condition, consider a (counterfactual) situation in which the behaviour described by the generalization in question is undisturbed by external factors, in the sense that these factors take values in the redundancy range. Call such a situation a 'comparative ceteris paribus situation'. Following Schurz (2002), I distinguish (a) a comparative ceteris paribus situation (in which other causal factors are held fixed), and (b) a situation in which a system is isolated from disturbing factors. Isolation is used in the sense that other causal influences are absent in an isolation situation (this conforms to the common usage of 'isolation' in physics). Instead, I suppose that other causal influences are held fixed.[15] Equipped with this terminology, one can refine minimal invariance without reference to interventions (as Woodward and Hitchcock understand the concept). I call the refined concept 'minimal invariance*'

A statement G of the form $Y = f(X)$ is minimally invariant* iff there are two values of Y that counterfactually depend on distinct values of $X$ in a comparative ceteris paribus situation; that is, a situation in which other causes take values in the redundancy range.

It is a great advantage to be able to define minimal invariance* purely in terms of counterfactual dependence in situations in which other factors are held fixed,[16] because – as argued in Chapter 4 – there are good reasons to believe that the notion of an intervention is itself problematic.

What is the pay-off of invariance theory? Invariance theory does not require lawish statements to be universal$_4$. Thus, adopting invariance theory seems to be one plausible way to account for the lawishness of generalizations that are non-universal$_4$. Furthermore, invariance theories are a promising tool with which to meet the requirement of relevance.[17]

## Meeting the challenges: Lange's Dilemma and the requirement of relevance

How does the distinction of four dimensions of non-universality help to meet the challenges? Accepting universality in the first and second dimension (i.e., we may say that special science laws are system laws), the theories of the third and the fourth dimension solve the problems. How? Lange's Dilemma addresses a problem of generalizations that are context-sensitive with respect to disturbing factors. Naturally, I claim that our theory of the third dimension has to deal with Lange's Dilemma: quasi-Newtonian laws play this role because they are describing the influence of disturbing factors. The requirement of relevance addresses a problem concerning the relation between the antecedent and consequent of the law statement. In terms of variables, the requirement of relevance addresses a problem concerning the question as to whether a law statement holds for different possible values of the antecedent variable(s). Naturally, I claim that our theory of the fourth dimension has to deal with the requirement of relevance: relevance is understood as invariance*. I will argue first that Lange's Dilemma can be avoided and, second, that the requirement of relevance can be fulfilled.

First, does adopting the view that lawish statements in the special sciences are quasi-Newtonian laws avoid Lange's Dilemma? I argue that it does: quasi-Newtonian laws describe the influence of disturbances. Assuming that laws are quasi-Newtonian avoids falsity because the occurrence of a disturbance does not render the law L in question false – instead, the influence of a disturbing factor is described by another law L* (and L* might be known or unknown by scientists). It avoids triviality because it is not (exclusively) committed to the fatal expression 'if nothing interferes' – rather, quasi-Newtonian laws describe two kinds of situation: undisturbed (i.e. 'inertial') behaviour and disturbed ('deviant') behaviour of a kind of system. It was argued that an advocate of the quasi-Newtonian approach is committed to the following existential claim: there are known and unknown disturbing factors, and known and unknown corresponding laws of deviation from the point of view of a special science.

Second, according to Earman and Roberts (1999), explaining why a counter-instance to L occurred is not enough to avoid Lange's Dilemma – what has to be supplemented is an account of relevance. I called this the requirement of relevance. I think that being invariant* meets the challenge neatly, because relevance is spelled out in terms of counterfactual dependence between the possible values of those variables figuring in L. In other words, the antecedent A of a law L is relevant for the consequent C, if L is invariant*. (L holds for at least one alteration of A in a comparative ceteris paribus situation, in which other factors – i.e. other factors than those explicitly figuring in L itself – are held constant.)

Recall Earman and Roberts's example of a spurious generalization in which the antecedent is irrelevant for the consequent: 'all spherical bodies conduct electricity' (Earman and Roberts 1999: 253). According to Hitchcock and Woodward's definition of minimal invariance, it might as well be that, in fact, all actual spherical bodies conduct electricity – however, the correlation of being spherical and conductivity turns out to be accidental, because one can at least imagine a counterfactual situation in which the geometrical shape of the body in question is changed from spherical to cubical by intervention and, yet, the conductivity remains unchanged. Analogously, my refined notion of invariance* assumes that the counterfactual 'if it were the case that (a) the geometrical shape of the body were cubical (and not spherical, as it actually is), and (b) the body were in a comparative ceteris paribus situation, then the body would conduct electricity differently than it actually does' is false. This evaluation of the counterfactual is, of course, in accord with Earman and Roberts's intuitions. Thus, both versions of invariance theories correctly classify the generalization 'all spherical bodies conduct electricity' as accidental, because counterfactual changes of the antecedent are irrelevant for a change of the consequent. In other words, being spherical fails to be relevant for conductivity of electricity, because the statement 'all spherical bodies conduct electricity' fails to be minimally invariant.[18]

To sum up, quasi-Newtonian laws (i.e. a non-epistemic, objective, ontological version of Pietroski and Rey's ECC) and an account of relevance in terms of invariance* are supplements for a theory of lawish statements in the special sciences.

## Conclusion: the explication of special science laws

The chapter began by asking how non-universal generalizations in the special sciences can perform a lawish function. Four dimensions of universality were distinguished. Further, the development of a theory was demanded for each dimension, which explains why special science generalizations can play a lawish role. It was argued that the universality in the first and second dimension for laws of the special sciences should be preserved – taking this into account is still compatible with describing special science laws as system laws (cf. Schurz 2002). The non-universality in the third dimension is taken care of by the Mill–Maudlin view of quasi-Newtonian laws. We deal with the fourth dimension of non-universality by relying on the refined notions of minimal invariance*. Based on these results, my explication of a special science law is the following:

> A statement L is a special science law iff (1) L is a system law, (2) L is quasi-Newtonian, and (3) L is minimally invariant*.

It was argued that this explication has benefits: (a) it allows the avoidance of Lange's Dilemma, and, (b) it meets the requirement of relevance in terms of invariance. For these two reasons, it was concluded that, according to my reconstruction, non-universal special science laws are, at least, good candidates for true, empirical (i.e. not trivially true) statements playing a lawish role in the special sciences. (c) For the dialectic of the book, this explication is advantageous because, first, it does not make reference to interventions (see Chapter 4 for arguments against interventions), and, second, this explication of laws is employed in my own explication of causation in the special sciences, the comparative variability theory (see Chapter 8).

# 6

# Woodward Meets Russell: Does Causation Fit into the World of Physics?

## Non-reductive explications of causation and metaphysics

In the previous chapters, objections to Woodward's interventionist theory of causation consisted in arguing that interventions are troubled because they are either dispensable for the semantic project with respect to causation, or they lead to an inappropriate evaluation of interventionist counterfactuals, at least if the antecedent of such a conditional involves an existential claim about merely logically possible interventions (see Chapter 4). The results of Chapter 4 lead to a follow-up problem for interventionists: the invariance account of laws in the special sciences – which interventionists endorse – cannot be maintained because this account uses the notion of a possible intervention (see Chapter 5). An alternative explication of laws in the special sciences was offered that does not suffer from dependence on possible interventions. This chapter will present another objection to the use of the notion of an intervention. Interventionists argue, as will be shown, that the notion of an intervention is of great value because it allows us to explain why causation has certain features, such as being asymmetric and time-asymmetric. The interventionists' positive argument for the theoretical indispensability and value of interventions is what I term the 'open-systems argument'. This argument, which is crucially built on the notion of an intervention, is supposed to explain why causation has certain features. I take this to be a metaphysical claim. As argued in Chapter 2 (pp. 54–5), interventionists are in need of an argument that explains why causation is an asymmetric and time-asymmetric relation, because it does not follow from the interventionist theory alone that causation has these features. (Recall that the time-asymmetry of causation implies the causal asymmetry, but not vice versa. For this reason, explaining the time-asymmetry of causation also provides an indirect explanation of the causal asymmetry.) This poses a challenge for interventionists, because the naturalist criterion of adequacy requires that an adequate theory of causation in the special sciences ought to explain why causation possesses certain characterizing features (see Chapter 1, pp. 15–21).

In this chapter, it will be argued that the interventionists' open-system argument is not sound. Consequently, interventionists cannot explain why causation has the features of being asymmetric and time-asymmetric. If the open-systems argument is not sound, interventionists also fail to provide a positive reason for the claim that interventions are indispensable for the metaphysics of causation (as well as for an account of the truth conditions of causal claims). However, an alternative explanation will be presented that accounts for the characterizing features of causation (especially for asymmetry and time-asymmetry). It is an advantage of the alternative explanation that it is able support my comparative variability theory (Chapter 8).

Let me begin by presenting a metaphysical problem for the interventionist theory of causation and other conceptually non-reductive theories of causation. In the recent debate on causation, several attempts of conceptually non-reductive explications of causation have been proposed. These explications are non-reductive in the sense that a causal notion is explicated by means of (other) causal notions. Prominent examples of the conceptually non-reductive approach are the following ones: most importantly for my purposes, Woodward's (2003: 59) interventionist theory of causation is conceptually non-reductive because it refers to interventions (which are explicitly introduced as causal notions) and it includes a clause to hold other causes fixed (similar accounts are presented by Hausman 1998; Hitchcock 2001; Halpern and Pearl 2005); Cartwright's classic paper 'Causal Laws and Effective Strategies' advocates a probabilistic explication of causation that is non-reductive because it refers to 'causally homogenous situations' (1983: 25); Schaffer (2004b: 305) argues for a Lewisian counterfactual theory of causation; however, Schaffer's preferred semantics for counterfactuals is non-reductive because it employs a similarity relation in semantics that essentially depends on causal independence conditions (similarly Kment 2006).

What are the reasons to accept non-reductive theories of causation? In my opinion, there are two compelling reasons: first, they cope with alleged counter-examples extremely well (such as various scenarios of pre-emption, over-determination, and other scenarios that are subsumed under the distinction criterion of adequacy – see Chapter 1, pp. 54–70, and Chapter 7). Second, they seem to reconstruct paradigmatic examples of causal statements from the special sciences adequately in the sense that they conform to the naturalist criterion. Chapter 2 attempted to show that at least interventionist theories are adequate in this sense (although I diagnosed that some desiderata remain with respect to the no-universal-laws requirement, the causal asymmetry, and the time-asymmetry of causation). Despite these reasons in favour of conceptually non-reductive explications, one might still remain sceptical and wonder whether it is not the case that conceptual circularity is problematic because being circular prevents an explication from being informative in any way. In other words, the worry is that conceptually non-reductive explications of causation are trivial and, therefore, useless. Certainly, the threat of triviality is a serious challenge to the approaches mentioned here, and it is also a challenge to my own theory of causation, presented in Chapter 8. Chapter 7 will defend conceptually non-reductive accounts against the charge of vicious circularity. However convincing this defence of conceptually non-reductive explication might turn out to be, let us suppose for the sake of argument that this circularity is not vicious as the proponents of the non-reductive explication argue (cf. Woodward 2003: 104–7; Schaffer 2004b: 307f.). Assuming that non-reductive explications do not run into a conceptual problem (i.e. the threat of triviality), I wish to raise another problem: i.e. whether conceptual circularity leads to a problematic metaphysics of causation. This chapter will depart from the primary task of this book, the conceptual project with respect to causation in the special sciences, and turn to metaphysical questions, because circular definitions of causation suggest a metaphysics that is controversial in the philosophy of causation.

Let us take a step back in order to see why a metaphysical problem could possibly arise. Given a conceptually non-reductive explication of causation, what does such an explication suggest about the metaphysics of causation? Some philosophers might be tempted to think that a conceptually non-reductive explication of causal notions at least strongly suggests that causal facts are ontologically irreducible and fundamental (for instance, Cartwright 1983). This view certainly is a metaphysical option. However, many philosophers of science reject the view that causation is an irreducible, fundamental building block of the world. The main source of this critical attitude towards a metaphysics of fundamental causal facts is a tension between: (a) the assumption that causal facts are fundamental, (b) the assumption that the fundamental facts are described completely by the present best theories of physics (or, at least, the presently best theories of physics are believed to be good approximations of the behaviour of fundamental entities), and (c) as the sceptics of fundamental causation argue, several features of these physical theories preclude that causation can be a fundamental. The most influential advocate of this physics-based denial of the claim that causation can be fundamental is Bertrand Russell. In his paper 'On the Notion of Cause', Russell (1912/13) presents a number of arguments to the conclusion that there is no place for causation in the physical realm.

A rejoinder to this physics-based denial of fundamental causation consists in rejecting the premise implicitly underlying the claims (a)–(c) above that (some) physical theories (aim to) describe the *fundamental* features of the world. In other words, first, one rejects the claim that physics describes the fundamental level of the world, and, second, one maintains the philosophical claim that the world is fundamentally causal (cf. Tooley 2009). According to this view, physics is incomplete because it does not describe the fundamental causal structure of the world. However, although this approach is certainly an option concerning the metaphysics of causation, I think it rests on the unattractively strong claim that we are able to contest the empirical strength of physical theories (i.e., for instance, the completeness of a physical theory) based on purely (non-empirical) philosophical reasons. In other words, we ground our beliefs that there are causes on the fundamental level of the world in intuition-based, a priori arguments *regardless of* how physics describes the world at its fundamental level. Again, this strategy certainly is a philosophical option. A metaphysics of causation of this kind is deeply at odds with the naturalistic methodology which is presupposed in this book: the sciences ought to guide our ontological commitments and, in cases of conflict between the sciences and our everyday intuitions, we cannot simply reject what science tells us.

In this chapter, I will reconstruct two of Russell's arguments for the claim that causation is not fundamental according to physics. Further, I will present a more modest neo-Russellian view that I ascribe to contemporary Russellians: they claim that there are no causal facts on the fundamental level (orthodox Russellian claim), if this level is described by fundamental physics – however, there are non-fundamental causal facts of the special sciences (neo-Russellian claim). It is precisely this view I wish to defend. Several objections are raised to the open-systems argument for the neo-Russellian claim that is advocated by interventionists, and which relies essentially on the notion of a possible intervention.

It will be argued that even interventionists can rely on another argument for the neo-Russellian claim: the statistical-mechanical account of the metaphysics of causation (cf. Albert 2000; Kutach 2007; Loewer 2007, 2009). If this reasoning is sound, it can be concluded that interventionists fail to provide reasons for the indispensability of interventions. However, interventionist theories of causation and, in general, conceptually non-reductive theories of causation (that are successful in satisfying the naturalist and the distinction standard of adequacy) are not committed to fundamental causal facts, if they are willing to adopt the statistical mechanical account.

## Russell on causation and the neo-Russellian claim

In 'On the Notion of Cause' (1912/13), Russell's simple claim is that in the 'advanced sciences' researchers do not aim at discovering causes; neither is the word 'to cause' used (also cf. Mach 1980: 278). Note that there are two readings of what 'use' is supposed to mean: according to the first reading, Russell claims that physicists neither utter nor write down the word 'cause' or other causally impregnated notions (such as 'production'). Descriptively speaking, this claim is definitely false (cf. Suppes 1970: 5f.; Hitchcock 2007: 55f.). According to the other and more charitable reading, Russell intends to say that there is nothing in physics that satisfies the description of being a cause. Whether physicists use words such as 'cause' and 'causal' is of no real importance. What is important is whether denoting something as causal serves any substantial purpose – that is, whether causal notions used by physicists are, indeed, associated with the characteristic features that philosophers assign to causation (cf. Norton 2007b, 2009; Frisch 2009a, 2009b, for a discussion of this point).

'All philosophers, of every school' (Russell 1912/13: 1) nevertheless continue to analyse the notion of causation and to make metaphysical assumptions about causation (e.g. the claim that it is always the case that if the cause occurs, then the effect occurs). Russell endorses the naturalist principle that the 'advanced' sciences are our best guide to the ontology one should accept. Thus, philosophers are plainly wrong to assume that there are causes. Given the achievements of the contemporary advanced sciences, Russell argues that causes are 'the relic of a bygone age'. In other words, according to Russell, philosophers dealing with causation in the age of modern physics fail to be well-informed of the 'advanced sciences' of their days. At best, one could say in a Russellian spirit, the philosophical fans of causation are informed by the sciences of a bygone age.[1]

Is Russell's claim convincing? Russell's claim has recently been defended in philosophy of physics (cf. Redhead 1990: 146f.; Batterman 2002: 126; Norton 2007a: 21f., 2007b: 223; Ney 2009). However, various other neo-Russellian philosophers (Eagle 2007; Hitchcock 2007; Kutach 2007; Ladyman and Ross 2007; Loewer 2007; Ross and Spurrett 2007; Woodward 2007) have pointed out – or, at least one can understand their arguments in this way – that there is a problem arising from Russell's expression 'the advanced sciences'. The trouble is that being a naturalist with respect to the metaphysics of causation (in the sense that the sciences are the best guides for making ontological commitments concerning causation) does not lead to a unique ontological commitment. Even if it is true that physics does not commit us to the existence of (fundamental) causal facts, the special sciences (such as chemistry, biology, psychology, economics and, also, several non-fundamental branches of physics such as, e.g., optics) do search for causes, they do endorse causal claims, and they do test causal models. If this is an adequate picture of what researchers in the special sciences do on an everyday basis, then, certainly, the nature of inquiry in the special sciences does not fit Russell's characterization of an advanced science. And, to be sure, all of Russell's examples for advanced sciences are case studies from physics (one of Russell's main examples is the law of gravitation, cf. Russell 1912/13: 13f.). Yet, naturalists about the metaphysics of causation have no reason to doubt that the special sciences are as scientific as physics despite the differences of these disciplines. This naturalist stance leads neo-Russellians to observe that, to use a formulation of Hitchcock (2007):

- where Russell went *wrong* is to ignore that the special sciences seem to seek causes and there is no reason to doubt that, say, biology and economics are not 'advanced' scientific enterprises; and
- where Russell was *right* is that there are no causes according to fundamental physics.

These observations amount to what I call the *naturalist challenge for neo-Russellians*. The challenge is to answer the question: How can there be causes in the *special sciences* if there are no causes in *fundamental physics*? Before returning to the neo-Russellian challenge, I first turn to two arguments by Russell.

Let me reconstruct two of Russell's most important arguments for the claim that there is no causation in fundamental physics. In order to understand these arguments properly, it is important to acknowledge that Russell ascribes various characteristic properties to causal relations, and so, at least in part, do the neo-Russellians:

1. *Sufficiency*: Causes are sufficient for their effects; that is, it is always the case that if the cause occurs, then the effect occurs (cf. Russell 1912/13: 7–12). Causes 'compel' (cf. Russell 1912/13: 2, 9f.), enforce, or produce their effects; that is, some kind of modal connection obtains among cause and effect.
2. *Time-asymmetry*: Causes precede their effects in time, but not vice versa (cf. Russell 1912/13: 13–16).
3. *Causal asymmetry*: The causal relation is asymmetric; that is, if A causes B, then B does not cause A (cf. Russell 1912/13: 10).
4. *Locality*: Cause and effect are local and distinct events in space-time region $r$, where $r$ is 'something short of the whole state of the universe' (cf. Russell 1912/13: 7).

In the current debate about the metaphysics of causation, these features of causal relations are thought of as explicating the folk notion of causation (cf. Norton 2007a: 36–8; Ladyman and Ross 2007: 268f.; Ross and Spurrett 2007: 13f.). As these philosophers point out, causation is often characterized by these features not only in ordinary discourse, but also in special science discourse. In this respect, it is misleading to call the notion of causation a 'folk' notion. However, I will adopt the terms 'folk notion/features' of causation, as it is an established and useful term in the debate. Presupposing that these features (1)–(4) characterize causal relations, Russell argues that the relations of physics lack precisely these features, and, thus, relations in physics cannot be causal ones. Note that, in this chapter, the focus will be on the important features of causal asymmetry and time-asymmetry, because they are the most uncontroversial and most challenging characteristics of causation, and to account for these features remains to be desideratum for interventionists (see Chapter 2, pp. 54–5). Let us now turn to Russell's first argument.

### Russell's first argument

The first argument is directed against the feature of *sufficiency* of causal relations. These features are essential to a Millian regularity theory of causation, which was one of Russell's primary targets. Mill defines causation in terms of sufficiency as follows: 'To certain facts certain facts always do, and, as we believe, will continue to, succeed. The invariable antecedent is termed the cause; the invariable consequent the effect' (Mill 1891: 213). Russell objects to this regularity view of causation:

> In order to be sure of the expected effect, we must know that there is nothing in the environment to interfere with it. But this means that the supposed cause is not, by itself, adequate to insure the effect. And as soon as we include the environment, the probability of repetition is diminished, until at last, when the whole environment is included, the probability becomes *nil*. (Russell 1912/3: 7f.)

Russell argues that sufficiency cannot be true, because one has to acknowledge the actual occurrence of interfering factors. The Millian regularity theorist faces the difficulty that causes are not sufficient for their effects because some event (which is distinct from the cause) might interfere and prevent the occurrence of the effect. If one includes a list of actual interferences in the description of the cause (in the worst case scenario, a description of 'the whole environment' of cause and effect) in order to vindicate sufficiency, then one faces the following situation: either it is *unlikely* that events of the cause type, of the effect type, *and* of the *whole environment type* occur regularly (i.e. they do not instantiate universal regularities), or, in the worst case for regularity theorists, events of these kinds do *not* reoccur *at all*.

As noted, Russell's first argument attacks an idea that is typical for regularity theories (sufficiency). The first argument might be a striking objection to regularity theories of causation that are similar to Mill's account. However, the feature of sufficiency is not presupposed by the non-reductive probabilistic, counterfactual, and interventionist theories of causation that have been introduced (p. 147). Revisiting the debate on causation after Russell, it might even have been the case that, historically, Russellian arguments have convinced these philosophers to reject the claim that causes are sufficient for the occurrence of their effects.[2] According to these theories of causation, causes are neither required to be sufficient causes, nor are perfect regularities necessary for causation. Neither is it the case that my own comparative variability theory of causation (see Chapter 8) holds that sufficiency is a necessary condition of causation. Rather, these theories have already considered interference cases as described in Russell's first argument: the modal connection of causation is understood in terms of either probabilistic or counterfactual dependence, and not in terms of invariable succession or entailment by a proposition describing a perfect regularity. Further, the sensitivity of counterfactual and probabilistic dependence relations to disturbing factors is dealt with by, for instance, ceteris paribus clauses and invariance conditions (cf. Hitchcock 2001; Woodward 2003), a similarity ordering of worlds (cf. Lewis 1973a), and statistical independence principles such as Reichenbach's common cause principle (cf. Reichenbach 1956) and the causal Markov condition (cf. Pearl 2000; Spirtes et al. 2000). In other words, all of these theories claim that inferences from cause to effect are non-monotonic – that is, the inference from a statement $X = x$ about the occurrence of the cause to a statement about the occurrence of the effect $Y = y$ does not remain sound if we introduce *any* arbitrary premise $W = w$ (because $W = w$ may be a statement about the occurrence of a disturbing factor).

### Russell's second argument

Another famous argument against the existence of causation in fundamental physics is directed against the time-asymmetry of causal relations. In contrast to the rejection of sufficiency, it is the case that probabilistic, counterfactual, and interventionist theories, and my own approach accept that causal relations are *time-asymmetric*. Moreover, each of these theories aims at explaining why causation has the feature of time-asymmetry. Concerning the time-asymmetry of causation, Russell points out a fact about fundamental equations in physics that he illustrates with the law of gravitation:

> The law [of gravitation] makes no difference between past and future: the future 'determines' the past in exactly the same sense in which the past 'determines' the future. The word 'determine', here, has a purely logical significance: a certain number of variables 'determine' another variable if that variable is a function of them. (Russell 1912/3: 15)

The law of gravitation is Russell's paradigmatic example of 'what occurs in any advanced science' (Russell 1912/13: 14). He observes (1912/13: 15, 18, 22) that the fundamental laws of physics (such as the law of gravitation, Newton's laws of motion and – today a Russellian might add – the field equations of general relativity theory, the Schrödinger equation in quantum mechanics and so on) are time-*symmetric*. One common way to understand time-symmetry is time reversal invariance: the fundamental laws are time reversal invariant iff if the laws hold for a sequence $ of states of a physical system $S_1(t_1)$, ..., $S_n(t_n)$, then they also hold for the temporally reversed sequence $* of states $S_n(t_n)$, ..., $S_1(t_1)$.[3] The fundamental physical laws state time-symmetric (or, to use an alternative name for this feature, time-reversible invariant) functional relations among the states of physical systems. Russell and the neo-Russellians point out that the time-symmetry of the fundamental laws of physics flatly contradicts the claim that there are fundamental physical time-asymmetric causal relations. Time-*asymmetry* is an essential feature of causal relations according to the folk notion of causation (cf. Ladyman and Ross 2007: 268f.; Norton 2007a: 36–8; Ross and Spurrett 2007: 13f.) and the naturalist criterion of adequacy. Thus, Russell and the neo-Russellians conclude that there are no causal relations among the fundamental physical facts.

On my opinion, the second argument is a serious challenge for the theories of causation mentioned at the beginning of this chapter. I think a serious challenge arises for two connected reasons.

First, the beginning of the chapter considered a metaphysical option for proponents of an interventionist theory of causation: one might be tempted to think that a conceptually non-reductive explication of causal notions at least strongly suggests that causal facts are ontologically irreducible and fundamental. At least, Russell's second argument should be a good reason for interventionists to cast doubt on the metaphysical claim of causal fundamentalism and to give causal anti-fundamentalism a second look. So, is there an alternative to fundamental causal facts for interventionists? An attractive alternative metaphysical option seems to be the neo-Russellian account, which falls into two theses:

- *Neo-Russellian claim*: There are higher-level causal facts.
- *Orthodox Russellian claim*: There are no fundamental causal facts.

Second, if one opts for the neo-Russellian account, then one has to tell a coherent story of the following three claims:[4]

- *Neo-Russellian claim*: There are higher-level causal facts.
- *Orthodox Russellian claim*: There are no fundamental causal facts.
- *Dependence claim*: Higher-level causal facts metaphysically depend on fundamental physical facts.

The additional third constraint is important because it prevents taking the easy way out: the constraint precludes the option simply to accept both (a) non-causal fundamental physical facts, and (b) causal facts in the 'higher levels' of the special sciences. Such an 'easy way out' view would, for instance, treat higher-level causal facts as strongly emergent from acausal fundamental physical facts. Many philosophers believe that physics plays a special role, and that this role constrains the ontology of other sciences. In the debate, philosophers usually endorse the dependence claim in terms of supervenience. That is, acausal physical facts are the supervenience base for special science facts, including the causal facts of the special sciences. The resulting task for neo-Russellians is to provide a coherent physical explanation of why the conjunction of the neo-Russellian claim, the orthodox Russellian claim, and the dependence claim is true. If such an explanation can be provided, then higher-level causal facts are physically kosher facts and the neo-Russellian challenge is met. This strategy consists in providing an account of an – at least, in principle – physical explanation of why certain causal special science facts obtain, although there is no causation in fundamental physics. The following two sections will discuss two strategies to meet the neo-Russellian challenge. The interventionists' open-systems argument will be discussed, and it will be argued that this argument is not sound. The chapter goes on to advocate a statistical-mechanical approach.

## The interventionist open-systems argument

Several proponents of interventionist theories of causation wish to meet the neo-Russellian challenge. These philosophers all (surprisingly) agree on an argument – the open-systems argument – which is essentially based on the notion of an intervention and the distinction between open and closed systems (cf. Eagle 2007: 171; Hitchcock 2007: 53f.; Woodward 2007: 92f.). Further, they are all inspired by a claim by Judea Pearl:

> *If you wish to include the entire universe in the model, causality disappears because interventions disappear* – the manipulator and the manipulated lose their distinction. However scientists rarely consider the entirety of the universe as an object of investigation. In most cases the scientist carves a piece from the universe and proclaims that piece *in* – namely the *focus* of investigation. The rest of the universe is then considered *out* or *background* and is summarized by what we call *boundary conditions. This choice of ins and outs creates asymmetry in the way we look at things, and it is this asymmetry that permits us to talk about 'outside intervention' and hence about causality and cause-effect directionality.* (Pearl 2000: 349f, emphasis added)

The explanatory targets in this quote seem to be features of causality, most importantly the feature of asymmetry. And the notion of an 'outside intervention' seems to play a key role in this explanation. What is the argument captured in this brief quote by Pearl? Let us focus on Woodward's argument because his is the most elaborated; furthermore, Hitckcock's and Eagle's approaches appear to be fairly similar.

Before I turn to Woodward's version of the open-systems argument, I will briefly add two comments regarding Woodward's stance on the neo-Russellian claim.

First, Woodward deals primarily with the pragmatic issue of whether causal notions, explicated in terms of his own interventionist theory of causation, are applicable to fundamental physics in the same way as they are, in fact, successfully applied in contexts of the special sciences, juridical law, and everyday life (cf. Woodward 2007: 67, 70, 93, 102). Woodward arrives at a negative conclusion: causal notions that are explicated by an interventionist theory of causation cannot be applied to paradigmatic cases in fundamental physics. Woodward is not willing to conclude from this result that causal notions do not play any role in fundamental physics. He suggests that, in fundamental physics, another concept of causation and, more generally, a notion of difference-making is used – the latter, Woodward seems to suggest, is not the primary target of his own theory of causation. Whether or not the latter claim about fundamental physics is true, the failure to apply causal notions – understood along the lines of interventionism – to fundamental physics amounts to the truth of the orthodox Russellian claim. The successful application of causal concepts in contexts of the special sciences, juridical law, and everyday life amounts to the neo-Russellian claim (although it is formulated in a modestly pragmatic rhetoric of the applicability of causal concepts).

Second, Woodward is very careful not to simplify the epistemic and metaphysical relations between causal claims in the special sciences and the fundamental physical laws. Woodward (2007: 80–4) argues, to the best of my understanding, that we should not expect these relations to be always as neat as in the case of the theoretical reduction of thermodynamics to statistical mechanics, because special science predicates are often more 'coarse-grained' than the predicates of thermodynamics, and special sciences laws backing up causal statements in these disciplines are often less stable and less invariant than the laws of thermodynamics. However, even if these remarks were true, they do not contradict the claim that physical facts are the supervenience base for special science facts, including causal facts of the special sciences. Indeed, Woodward (2007: 99–101) uses an analogy in order to establish the objectivity of special science causal facts: the analogy between objective chances in systems that are governed by deterministic laws (Woodward refers to Strevens 2003) and the objectivity of special science causation in a world that is governed by non-causal fundamental physical laws. The positive analogy is that objective chances and causal relations are non-fundamental – however, both are perfectly objective macro-facts. However, the analogy is also in support of the supervenience claim that special science macro-facts (concerning causation and objective probability) are grounded in fundamental physical facts. The only amendment Woodward suggests is that special science facts are grounded not only in fundamental physical laws, but also in other particular physical facts (most importantly, initial and boundary conditions of a system). I will return to these remarks by Woodward, because they indicate an alternative argument to the open-systems argument that Woodward and other interventionists advocate.

I will now turn to Woodward's version of the open-systems argument. Woodward observes that the special sciences typically not only use causal vocabulary, but also that the systems they describe by means of causal claims exhibit a common feature: these special science systems typically are a small part of the entire world (as indicated in the quote by Pearl). By contrast, Woodward thinks that the systems described by physics are global as physical models describe states of the entire world (see Frisch (2010), for a critique of this view of theories in physics).[5] Woodward argues that the successful application of causal notions is explained by the fact that the special science systems are 'non-global', 'small', or 'open' (cf. Woodward 2007: 91):

> This has to do with the fact that such systems are typically only a small part of a much larger world or the environment which is outside the scope of the inquirer's interest *but which can serve as source of interventions.* (Woodward 2007: 90, emphasis added)

Woodward continues to point out that the existence of possible interventions (as presupposed in his theory of causation) requires open systems:

> One reason why interventionist counterfactuals seem unproblematic in this case is that we are dealing with [...] a system [...] that is isolated enough from its environment that it can serve as a distinctive subject of causal inquiry but not so isolated (or 'closed') that the idea of outside influence in the form of interventions makes doubtful sense. Put slightly differently, *the system of interest is located in a larger environment which serves as a potential source of 'exogenous' interventions.* (Woodward 2007: 92)

So, Woodward infers, if interventions require a 'larger environment' outside of a system that is subject to the intended application of a causal model, then including the whole universe in a model (as he supposes physicists do) leaves no room for interventions. In other words, there is no room for the 'potential source' of interventions. Given the interventionist theory of causation, Woodward argues, causal notions are not applicable to models that describe global and isolated systems; for example, models that describe the entire universe. Woodward illustrates his argument as follows:

> consider the claim that (U) the state $S_t$ of the entire universe at time t causes the state $S_{t+d}$ of the entire universe at time $t+d$. On an interventionist construal, this claim would be unpacked as a claim to the effect that under some possible intervention that changes $S_t$, there would be an associate change in $S_{t+d}$. The obvious worry is that it is unclear what would be involved in such an intervention and unclear how to assess what would happen if it were to occur, given the stipulation that $S_t$ is a specification of the entire state of the universe. (Woodward 2007: 93)

Putting all the pieces together, Woodward's open-systems argument works as follows:

### The open-systems argument

1. The interventionist theory of causation is true.
2. If the interventionist theory of causation is true, then causal claims may be true or false only of open systems – that is, systems that can be subject to a possible intervention from the outside, the environment of the system.
3. The special sciences typically deal with open systems.
4. Fundamental physics typically deals with closed systems (ultimately, the entire universe is a physical system exhibiting these features).
5. *Therefore*, causal claims may be true or false only with respect to open systems but not with respect to closed systems.

Woodward seems to argue that the open-systems argument meets the neo-Russellian challenge since it supports:

- The neo-Russellian claim that there are higher-level causal facts of the special sciences (the sciences of open systems); and
- the orthodox Russellian claim that there is no fundamental causation according to fundamental physics (the science of closed systems).

The question arises whether the interventionist open-systems argument is a convincing argument, which I do not think it is, based on the following four reasons.

*First objection*

Interventionists take it, at least, as a contingent fact (or even merely stipulate) that interventions on an open system influence only the future of the open system and not its past (cf. Frisch 2007: 355). This is no surprise since interventions are (especially constrained) causes. Interventions qua causes contingently occur before their effects. However, this contingent fact – that is, that interventions qua causes contingently influence the future of the system that is subject to the intervention – is the explanandum for neo-Russellians. It is the ambition of neo-Russellians to explain why the influence of all kinds of causes (including interventions as a special kind of cause) is contingently directed from the past to the future and not vice versa. Therefore, relying on interventions fails to explain why there is higher-level causation in a fundamentally acausal world (cf. Reichenbach 1956: 45; Mackie 1965: 51). This is precisely the challenge that Woodward, as a neo-Russellian, ought to meet.

*Second objection*

It is not clear to me why interventions on closed systems are supposed to be impossible. The claim that interventions of this kind are supposed to be impossible, if one recalls Woodward's definition of an intervention (Woodward 2003: 98f.) and his additional assumption that interventions are required to be merely logically – but not physically – possible (cf. Woodward 2003: 128, 132; 2007: 91). If one accepts this definition and his assumption about the modal character of interventions, I fail to realize why an intervention on a closed system of fundamental physics is supposed to be impossible. Imagine a universe in which the motion of particles is governed by Newtonian laws of motion (cf. Woodward 2007: 93). The laws of this universe are deterministic, global and complete in the sense that, given some state of the entire universe $S_1(t_1)$, the laws determine the state of the entire universe at any other past and future time. Let us assume that a state of the entire universe $S_1$ at $t_1$ is the cause of a distinct entire state of the universe $S_2$ at a later time $t_2$. According to the interventionist definition of causation, $S_1(t_1)$ is a cause of $S_2(t_2)$ iff there is a possible intervention on $S_1(t_1)$ that changes $S_2(t_2)$. To claim that a possible intervention on $S_1(t_1)$ exists means that a variable modelling a magnitude in the Newtonian universe is changed because of the occurrence of an event *i* (that satisfies the definition of an intervention). In the framework of possible worlds semantics, one might state it like this: there is a logically possible world w in which $S_1(t_1)$ is the case, $S_1(t_1)$ is the result of an intervention, and w is a Newtonian world. What could such an intervention look like? For instance, the velocity of a particle is changed by the event *i* in a possible Newtonian world w. I see no reason why such an intervention should be *logically* impossible – and that is all Woodward requires. This world w is – apart from intervention event *i* – very similar to the actual Newtonian universe. And we seem to have a good reason to evaluate the interventionist counterfactual 'if it were the case that $S_1^*(t_1)$ (which results from the intervention event *i*), then $S_2^*(t_2)$ would be the case' in a world such as w. The counterfactual in question is true if there is intervention-world such as w, in which $S_1^*(t_1)$ is the result of an intervention, and this world evolves into $S_2^*$ at $t_2$.

One might worry that referring to possible world semantics in order to understand Woodward's account of causation is wrong-headed, because Woodward explicitly rejects Lewis's possible world semantics for counterfactuals (cf. Woodward 2003: 133–45). However, recall from Chapter 3 that taking a closer look reveals that Woodward merely objects to Lewis's similarity measure for the closeness of worlds to a world w in which counterfactual is evaluated (cf. Woodward 2003: 139, 142, for two counter-examples against Lewis's similarity metric). Woodward's position is entirely coherent with the general idea of Lewisian semantics for counterfactuals; that is, a counterfactual is true at a world w iff the consequent is true in the closest antecedent-worlds.[6] This general idea of Lewis's semantics can be distinguished from Lewis's specific proposal for selecting the closest worlds (i.e. his similarity metric). One appealing way to understand Woodward is to say that he deviates from Lewis by using a different measure for closeness of worlds. From Woodward's point of view, the most obvious candidate for such a measure is this: the closest antecedent-worlds are those in which the antecedent is the outcome of an intervention (cf. Woodward 2003: 135f.; Woodward and Hitchcock 2003: 13f., for the claim that interventions play a similar role as Lewisian small miracles by selecting the closest possible antecedent-worlds). The main point to stress here is that using standard possible worlds semantics in order to understand (a) existential claims about possible interventions, and (b) interventionist counterfactuals is compatible with Woodward's theory of causation (including his objections to Lewis's similarity metric).[7]

A proponent of the open-systems argument could bring up the following rejoinder to my counter-argument: it is a matter of conceptual or metaphysical necessity that the actual universe is closed in the sense that it is essentially constituted by the entities of which it actually consists. So, if one added an intervention event $i$ to the actual universe it would no longer be the same entity. By virtue of conceptual or metaphysical necessity, there is no intervention on the entity called 'the actual universe', because intervention-worlds would constitute another object that is not identical with the actual universe.[8] However, even if this were the case, one could still maintain that a counterfactual universe w that is in the state $S_1^*$ at $t_1$ is similar enough to the actual universe (though a different entity) in order to evaluate the counterfactual 'if it were the case that $S_1^*(t_1)$, then $S_2^*(t_2)$ would be the case'. Moreover, one has good reason to believe that this counterfactual is true in the actual world, if the actual world is Newtonian, as assumed in the example.[9]

Does this show that Woodward was mistaken and there is, in fact, at least causation in closed systems if interventions on closed systems are possible? The third objection suggests that this is not the case.

*Third objection*

The second objection was that there is not a good reason to believe that it is (logically) *im*possible to intervene on a closed system. If this objection is correct, does it imply that causal claims are true with respect to closed systems such as Woodward's Newtonian universe? I will argue that this is not so, because the possibility to intervene simpliciter does not establish the time-asymmetry of causation. This claim might be surprising and requires clarification because – as observed in the first objection – interventions qua causes have a future-directed time-asymmetric influence as a matter of empirical fact. However, if interventionists want to avoid the first objection, then they should not require that interventions could merely influence the future of the system that is the subject of the intervention. I suggest that, in order to avoid the first objection, interventionists have to allow (a) synchronic interventions, or (b) backwards interventions. A synchronic intervention on a variable $X$ influences $X$ without any time-delay. A backwards intervention on $X$ is a special case of backward causation such that $X$ occurs earlier than the intervention. My concern about this strategy of avoiding the first objection is that, if interventionists accept synchronic or backwards interventions in order to respond to the first objection, then interventions do not help to establish the time-asymmetry of causation.

My argument is analogous to an argument by Adam Elga (2001) against David Lewis's semantics for counterfactuals. Elga argues against the Lewisian claim that the time-asymmetry of counterfactual dependence can be justified if one selects the closest antecedent-worlds by virtue of being small miracle worlds. Suppose event $a$ actually occurs at $t_1$ and event $c$ actually occurs at a later time $t_2$. Now, consider two counterfactuals: the non-backtracking counterfactual 'if non-$a$ were the case at $t_1$, then non-$c$ would be the case at a later time $t_2$' and the backtracking counterfactual 'if non-$c$ at $t_2$ were the case, then non-$a$ would be the case at $t_1$'. Elga claims that if one selects the closest antecedent-worlds of both conditionals by virtue of being small miracle worlds, both counterfactuals are evaluated as true. What is the argument supporting this claim? The argument rests on the premise that is shared by neo-Russellians: the dynamical fundamental laws of physics are time-symmetric. Elga argues that the closest antecedent-worlds for evaluating the non-backtracking counterfactual are: a small miracle occurs shortly before the antecedent-event non-$a$. The non-backtracking counterfactual is true if the small miracle leading to non-$a$ at $t_1$ leads to non-$c$ occurs at $t_2$ (instead of the actual event $c$) by running the time-symmetric dynamical fundamental laws forward in time. This is just the kind of result Lewis desires. However, Elga shows that a backtracking counterfactual is also evaluated as true in a small miracle world. According to Elga, the small miracle worlds in which backtracking counterfactuals are evaluated are such that the small miracle occurs shortly after the consequent-event $c$ – at $t_3$. The backtracking counterfactual is true if – by running the time-symmetric laws backwards in time from the miracle at $t_3$ on – the resulting course of events in the past of non-$c$ differs from the actual past of $c$; that is, non-$a$ is the case at $t_1$.[10] Hence, the backtracking counterfactual is true. Elga draws the conclusion that 'in this case there is no asymmetry of miracles, and hence in this case Lewis's analysis fails to yield the asymmetry of counterfactual dependence' (Elga 2001: 321f.).

An analogous argument can be directed against the attempt to establish the time-asymmetry of counterfactuals and – since Woodward's theory of causation is linked to interventionist counterfactuals – causation by relying on interventions. Consider Woodward's own example of a Newtonian universe (see second objection) to realize why this is the case. Suppose that $S_1{}^*(t_1)$ is a counterfactual state of the universe that is the effect of a future directed intervention that occurred at earlier time $t_0$ (and assume further to hold the past of the universe prior to the intervention at $t_0$ fixed). Accordingly, the counterfactual 'if it were the case that $S_1{}^*(t_1)$, then $S_2{}^*(t_2)$ would be the case' is true if, in the closest worlds, $S_1{}^*(t_1)$ is the result of the future directed intervention and – by running the laws forward in time – the universe evolves into the state

$S_2{}^*$ at $t_2$. The reason why we consider this counterfactual to be true is firmly tied to the assumption that interventions can only influence the future of Woodward's universe.

However, if we drop the assumption the interventions may only have a future-directed time-asymmetric influence – in order to avoid the first objection – the picture drastically changes. Analogously to Elga's argument and in accord with Woodward's example, the fundamental dynamical laws governing the Newtonian universe are time-symmetric. Once we introduce synchronic or backwards interventions, the time-symmetric laws governing Woodward's universe allow that the following backtracking counterfactual is true: 'if it were the case that $S_2{}^*(t_2)$, then $S_1{}^*(t_1)$ would be the case'.[11] Under which conditions is this conditional true? In analogy to Elga's argument, let us assume that $S_2{}^*(t_2)$ is the result of either a synchronic intervention at $t_2$ or of a backwards intervention at $t_3$ (and we assume that the future of the universe later than $t_2$, or respectively at $t_3$, is fixed). Then, we run the time-symmetric laws governing Woodward's Newtonian universe backwards from state $S_2{}^*$ at $t_2$ to the earlier state of the universe at $t_1$. The backtracking counterfactual is true if the state at $t_1$ is $S_1{}^*$. We have reason to believe that this is the case: if we do not already build a time-direction into the interventions, then we ought to accept that the counterfactual state $S_2{}^*(t_2)$ evolves – by running the time-symmetric dynamical laws backwards in time – into a counterfactual state $S_1{}^*(t_1)$ that differs from the actual state $S_1$ at $t_1$.

The upshot is that if the analogy to Elga's argument holds, then – contrary to the interventionist theory – it is possible to intervene on a closed system (such as Woodward's Newtonian universe) and it is not the case that time-asymmetric causal facts obtain with respect to this system. Therefore, the mere possibility to intervene is insufficient for supporting the neo-Russellian claim and the orthodox Russellian claim.

*Fourth objection*

It is worth noticing that – in the quote presented at the beginning of this section – Pearl is speaking from a methodological point of view. Pearl's project is to develop algorithms that allow us to infer causal models from statistical data. In contrast to this methodological project, the interventionist theories of causation pursue a semantic project: they aim at clarifying the truth conditions of causal statements. If one cares about the evidence for a causal statement and the methodology of inferring causal models, then the claim gains plausibility that the investigator has to be outside of the system that the causal statement of question is about. Especially, if intervening is an important part of one methodology of constructing causal models (as in Pearl's case), then Pearl's claim that 'causality disappears because interventions disappear – the manipulator and the manipulated lose their distinction' (Pearl 2000: 350) makes some sense. However, it is puzzling that Pearl adopts the open-systems argument, because his notion of an intervention, unlike Woodward's, does not require that interventions are exogenous causes and, hence, does not require an environment as a source for possible interventions. Even if Pearl were justified in being concerned that one cannot infer a causal model representing a closed system by means of intervening in these systems, it does not follow that this concern is also justified for the interventionist semantic project.

Summing up this section, I conclude that the open-systems argument is not a sound way for interventionists to meet the neo-Russellian challenge. The next section examines a potential alternative for interventionists.

## The statistical-mechanical argument

As argued in the previous section, the open-systems argument is not sound. There is an alternative option for arguing for the neo-Russellian claim (regardless of whether one is an interventionist or not). It is an interesting fact that the alternative argument is suggested by the final paragraph in Woodward's paper when he attempts to specify the physical supervenience base for causal facts in the special sciences:

> Typically, the grounds or truth-makers for upper-level causal claims like 'Cs cause Es' or 'particular event *c* caused particular event *e*' will involve many additional factors besides laws (and besides facts about whether *C*, *E*, *c* and *e* instantiate laws or are parts of conditions that instantiate laws etc.). *These additional factors will include very diffuse, messy, and non-local facts about initial and boundary conditions that do not obtain just as a matter of law and have little to do with whatever underlies or realizes C, E, c or e themselves.* (Woodward 2007: 103, emphasis added)

This brief concluding remark by Woodward fits a recent approach to the metaphysics of causation that philosophers of physics have presented as an argument for the neo-Russellian claim (Albert 2000; Kutach 2007; Loewer 2007, 2009).[12] In what follows I do not wish to commit myself to this approach, because several points about it are not entirely settled and some are clearly controversial (as, most prominently, Frisch 2007, 2010 argues). The important point regarding the statistical-mechanical approach for my purposes is that if one follows Albert, Kutach, and Loewer, then one is equipped with an argument that helps to meet the neo-Russellian challenge. (In principle, this argument is also useful for interventionists who do not wish to rely on the open-systems argument.)

The basic idea of the statistical-mechanical (SM) account is that features of causation (most importantly, its time-asymmetry) are physically kosher because they can be, in principle, explained by the fundamental time-symmetric laws plus additional assumptions. The SM-account can be reconstructed as an analogy argument: it assumes that the time-asymmetry of causation and other features of causation can be explained by the same set of premises that explains another, macro-physical time-asymmetric generalization, the second law of thermodynamics. In order to present the SM-account of the time-asymmetry of causation, let me first introduce the micro-physical explanation of the time-asymmetry of the second law of thermodynamics ('the Second Law').

The SM-approach builds on an original idea of Ludwig Boltzmann: the explanation of the time-asymmetric Second Law is, most importantly, based the on the time-symmetric laws of statistical mechanics and the so-called Past-Hypothesis (PH). The Second Law is a paradigmatic example of a time-asymmetric special science law for which such an explanation is available and this explanation relies on time-symmetric fundamental laws of motion. A seminal formulation of the Second Law is:

> The total entropy of the world (or of any isolated subsystem of the world), in the course of any possible transformation, either keeps at the same value or goes up. (Albert 2000: 32)

Let me provide an example as an illustration of the Second Law.

> Place an iron bar over a flame for half an hour. Place another one in a freezer for the same duration. Remove them and place them against one another. Within a short time the hot one will 'lose its heat' to the cold one. The new combined two-bar system will settle to a new equilibrium, one intermediate between the cold and hot bar's original temperatures. Eventually the bars will together settle to roughly room temperature. (Callender 2011)

The behaviour of physical systems described by the Second Law is not time-symmetric. In the case of the heated and cooled iron bars, the Second Law does not describe a transition from the iron bars at room temperature at the later time $t_2$ to a state at the earlier time $t_1$ in which one iron bar is heated and the other iron bar is cooled. Thus, the Second Law is a time-*a*symmetric, physical macro-law. The Second Law differs from the time-symmetric fundamental laws in this respect.

According to the SM-account (cf. Albert 2000: 96; Kutach 2007: 329–31; Loewer 2007: 298–304, 2009: 156–8), the time-asymmetric Second Law can be derived from the following premises:

- (Laws) the time-symmetric, dynamical, and fundamental laws of statistical mechanics;
- (PH) a proposition that the initial macro-state of the universe was a state of low entropy; and
- (PROB) the assumption of a uniform probability distribution over the physically possible initial micro-states of the universe compatible with PH (i.e. the possible realizers of the initial macro-state referred to in PH).

To be more precise, according to the SM-account, these premises entail that it is highly probable (though not certain) that macroscopic systems evolve time-asymmetrically in accord with the Second Law. In other words, this explanation is a statistical and approximate explanation of the Second Law. At first glance, it might appear puzzling how a time-asymmetry can be explained by the means of time-symmetric fundamental laws. So, how does this explanation work? And, which premise does the main explanatory work? According to the fundamental laws, the sequence from original temperature (at $t_1$) to room temperature (at $t_2$) could be reversed in time. Naturally, the fundamental laws by themselves cannot explain the time-asymmetry of the macro-law. The crucial explanatory import is due to time-asymmetric boundary conditions – the existence of these boundary conditions is expressed by PH, and the uniform probability distribution (PROB) over the realizers of this macro-condition. As Kutach (2007: 334f.) puts it: 'The correct explanation of thermodynamic asymmetries comes by way of an *asymmetric boundary condition*. Specifically, we posit low entropy in the past and no similar constraint in the future.' This is a highly interesting result because (a) the SM-argument provides an explanation of the Second Law, and (b) the SM-argument reconciles the claim that there are time-symmetric laws on the fundamental level and the claim that there are time-asymmetric laws on the (physical) macro-level.

Now, the crucial question is whether the SM-argument can also be used for explaining the time-asymmetry of causation? The SM-account suggests that the answer is 'Yes' (Albert 2000: 128–30; Kutach 2007: 338–42; Loewer 2009: 160). The reason for this affirmative answer seems to be: suppose we want an account for the time-asymmetry of causation of a specific singular causal relation, say the fact that event *c* causes event *e*. And suppose that *c* and *e* are macroscopic events. Further, assume that one adopts a counterfactual theory of causation (as do Albert, Kutach, and Loewer). According to the simplest counterfactual theory, if *c* causes *e*, then two counterfactual conditionals have to be true: (1) 'if it were the case that *c*, then it would be the case that *e*', and (2) 'if it were the case that non-*c*, then it would be the case that non-*e*'. The truth of these counterfactuals requires that: (1) regarding the first conditional, there is, in principle, an SM-argument to the conclusion that the probability that $c(t_0)$ is followed by $e(t_1)$ is very high, and (2) regarding the second conditional, there is, in principle, an SM-argument to the conclusion that the probability that non-$c(t_0)$ is followed by non-$e(t_1)$ is very high. In these cases, the SM-arguments for the conclusion $e(t_1)$ and, respectively, non-$e(t_1)$ have the following premises:

- (FACT) a proposition that event *c* (respectively, non-*c*) at $t_0$ and the macro-conditon of the entire world at $t_0$ is $M(t_0)$ as a contingent macro-fact
- (LAWS)
- (PH)
- (PROB).

This is a Goodmanian way of stating truth conditions of counterfactuals in the SM-framework, because counterfactuals are nothing but condensed SM-arguments (see Chapter 2 for details concerning Goodmanian semantics). Albert (2000: 128–30) seems to prefer this reading, when he states the truth conditions of counterfactuals by deriving the consequent of the antecedent by 'normal procedures of inference' (the latter amount to what I called the actual availability or the existence of an SM-argument, cf. Albert 2000: 96, 129). At least a plausible interpretation of Loewer's (2007: 317) semantics is to understand it as a possible worlds semantics selecting the closest antecedent-world by the truth of (Fact), (Laws), (PH), and (PROB) in these worlds. This selection function differs from Lewis's original similarity heuristic (Loewer 2007: 321f.;

see Chapter 3 for a reconstruction of Lewis' semantics). Kutach (2007: 338–42) proposes a suppositional theory of the meaning of counterfactuals: the degree of the assertibility of a conditional equals the expected objective probability of the consequent conditional on the antecedent, (Fact), (Laws), (PH), and (PROB).

By contrast, given the premises of the PH-argument, the probability of the time-reversed sequence (i.e. $e(t_1)$ evolves into $c(t_0)$) is extremely low – although the occurrence of a time-reversed sequence is physically possible. Loewer (2009) extends the SM-account to the explanation of asymmetric special science laws. This is an exact analogy to the SM explanation of the Second Law. Any conclusion of an SM-argument is physically kosher. Thus, if there is an SM-argument for any causal fact, then causation is physically kosher. Another way to formulate the SM-account consists in saying that causal facts supervene on the history of the actual worlds, which is constrained by (Fact), (Laws), (PH), and (PROB). Note that this supervenience base for causal facts is kosher with respect to fundamental physics (i.e. it is not denied that the fundamental laws are time-symmetric). This metaphysical interpretation seems to fit the thought that Woodward (2007: 103) expressed in the last paragraph of his paper (cited at the beginning of this section).

Loewer points out that his thesis of 'grounding' the feature of time-asymmetry of special science causes and laws in the facts expressed by the premises of the SM-argument is a purely metaphysical thesis:

> It would be a very tall order to show that the dynamical laws and PROB imply a probabilistic version of Gresham's law (or any other special science law). *No one will ever produce a deduction of a special science law* since the special sciences are about entities and systems that are incredibly complicated from the perspective of physics and unlike the super Laplacian demon we don't have a translation manual that tells us which micro states realize which special science properties. *Nevertheless there is a good reason to think that if SS is a special science law then its lawfulness is derived from PROB and the dynamical laws.* (Loewer 2009: 160, emphasis added)

Although Loewer discusses the example of a special science law, he wants to make a more general point that also covers cases of special science causation. Loewer seems to say that there is, in principle, an SM-argument for any causal and nomic fact of the special sciences. Yet, this metaphysical claim does not imply that this is how one acquires the knowledge that a particular causal fact obtains; neither does this 'in principle' explanation correctly reflect the rules of application for causal notions (the latter might still be most adequately captured by an interventionist explication).

Let me clarify the dialectic at this point: the SM-argument supports the neo-Russellian claim. However, it does not show that the orthodox Russellian claim is true. The latter claim is established by Russellian arguments such as Russell's second argument (see pp. 154–5). The second premise of the SM-argument, (LAWS), states that the dynamical laws of fundamental physics are not causal laws (because they express time-symmetric dependence relations). In other words, the orthodox Russellian claim figures in the second premise of the SM-argument, because the dynamical laws of fundamental physics are characterized as non-causal. This premise in conjunction with (FACT), (PH) and (PROB) is intended to be an argument to the conclusion that higher-level causal facts (as stated by the special sciences) and higher-level nomic facts (as the second law of thermodynamics) have the feature of time-asymmetry.

To sum up, there are two lessons to be learned from the SM-account of the metaphysics of causation. *First,* the SM-account explains how the conjunction of the orthodox Russellian claim, the neo-Russellian claim, and the dependence claim can be true. A naturalistic metaphysician of causation should be perfectly happy with the neo-Russellian view, because it simultaneously conforms to the ontological commitments of fundamental physics *and* of the special sciences.

*Second,* although causal relations are local relations (in the sense that cause and effect are localized in space-time), the SM-account implies that they are extremely extrinsic relations. Take, for instance, the case of actual causation. If one relies on the SM-approach, then one is committed to the claim that causation is an extrinsic relation, because whether event $c$ causes $e$ at time $t$ depends on the events $c$ and $e$ *plus* the macro-state of the entire world at $t$, fundamental dynamical laws, PH, and PROB. In other words, whether $c$ causes $e$ depends on the history of the entire universe until $t$. Whether or not this is problematic, relying on the SM-approach is incompatible with the assumption that causation is something intrinsic to cause and effect. However, the same is true of many difference-making approaches to causation (such as regularity theories, probabilistic and counterfactual approaches). So, if its being extrinsic should turn out to be problematic, this problem would not exclusively affect the SM-approach.

Suppose that you are convinced by the SM-approach and its ability to support the neo-Russellian claim. If so, you might still be left with a concern:[13] we started out with conceptually non-reductive theories of causation, and, because of (neo-)Russellian arguments against causal fundamentalism, we ended up with a reductive metaphysics of causation (the SM-account according to which the orthodox Russellian claim is true). The SM-approach is metaphysically reductive because it grounds the causal facts of the special sciences in acausal facts such as (FACT), (LAWS), (PH), and (PROB). Does this result reveal that, ultimately and contrary to the first impression, interventionist theories (and other conceptually non-reductive approaches) are reductive approaches to causation? Does this lead to a problem for my own comparative variability theory, because it is – like Woodward's theory – conceptually non-reductive? Well, it depends. Now it is a good time to remember the distinction of the conceptual, the metaphysical, and the methodological projects concerning causation which were sketched out at the beginning of the book (cf. Mackie 1980: viii–ix; Dowe 2000: ch. 1). But will this help?

Woodward (2008: 193–6) insists that he does not pursue the metaphysical project. Rather, he argues, the interventionist theory is solely dedicated to the conceptual project: it explicates the notions of causation used in the sciences (and, partly, those of everyday discourse). Strevens replies to Woodward that it is hard to explain how the metaphysical and the conceptual project can be separated if one relies on a truth-conditional semantics:

> In modern times, such a project [i.e. Woodward's declared goal of providing an account of the meaning of causal statements] is invariably interpreted as aiming to provide truth conditions for the sentences or thoughts in question, and therefore as aiming to specify those representations' truthmakers. *It may look like semantics, but it is also a kind of metaphysics.* (Strevens 2008: 184, emphasis added)

I think Strevens makes a very good point. So, we have to wonder whether it is the case that the conceptual and the metaphysical project can no longer be distinguished once one adopts truth-conditional semantics? Is it the case that we can no longer consistently claim that the interventionist conceptually non-reductive explications are supplemented with the reductive metaphysics of the SM-account? In other words, are we back to the problem from which we started: is interventionism committed to fundamental, irreducible causation and, thus, troubled by (neo-)Russellian arguments?

There are several lines to take in replying to Strevens. However, due to space constraints my discussion of the Canberra plan will take the form of a suggestion rather than the form of a conclusive argument. I will focus on one: the Canberra plan might offer an attractive way out. According to the Canberra plan, one can draw a distinction between conceptual roles and role-fillers (cf. Chalmers 1996; Jackson 1998). Consider the concept 'president of the US'. The conceptual role of 'president of the US' consists in typical features of this office: this person lives in the White House, is the leader of the government and so on. The project of determining the conceptual role of 'president of the US' is an a priori enterprise. The conceptual role of this office picks out the office-holders in the actual world (say, at the time of writing, to keep things simple): B. Obama. Obama is the actual role-filler of the president-role. The project of determining the role-filler is empirical – as opposed to the a priori project of determining the conceptual president-role.

The distinction of role and role-filler suggests the following characterization of the interventionist explication of causal concepts. The interventionist theory of causation describes a particular feature of the folk role of causation: the feature that causal relations can typically be exploited for interventions. Other features of the folk role (e.g. being a time-asymmetric, asymmetric, local and so on) seem to be taken for granted in interventionist definitions. By analogy to the president-case, determining the actual role-filler of the causal role is an empirical project. If the SM-account is true, then the actual role-filler is the conjunction of facts such as (FACT), (LAWS), (PH), and (PROB). If interventionists describe a feature of the conceptual role of causation and the SM-account specifies the role-filler, then one can draw a distinction between conceptual analysis (or the explication) and metaphysics. The conceptual project is concerned with the conceptual role while metaphysics is concerned with the role-filler.

Is the strategy to adopt the Canberra plan in response to Strevens convincing? The distinction between role and role-filler seems to establish a distinction between the conceptual and metaphysical questions without having to reject truth-conditional semantics. This is just what interventionists need. A framework in which such a distinction can be maintained is the methodological basis or precondition for interventionist neo-Russellians: only under the assumption that conceptual analysis and metaphysics are separable can interventionists argue that conceptually non-reductive theories of causation are compatible with a metaphysics that does not take causation to be a fundamental feature of the world.

However, although the Canberra plan appears to be attractive, it suggests two amendments to Woodward's account of causation. First, this proposal has consequences for how interventionists describe their own theory: when they speak about providing an account of the 'meaning' of causal claims, they should restrict this speech to explicating a feature of the conceptual role of causation. This project has to be distinguished from specifying the role-filler by proving mind-independent truth conditions for causal claims. This requires a change in the way interventionists describe the goal of their project, because Woodward often seems to be occupied with an account of truth-conditional meaning (Woodward 2003: 7–9, 122–4; 2008: 193–6). However, Woodward could just accept that the relevant sense of 'meaning' he is interested in is an explication of the conceptual role. Second, and more importantly, it is a controversial matter whether the conceptual role of causation indeed picks out a role-filler as demanded by the Canberra plan. The Canberra plan faces two major problems when applied to causation: (a) the conceptual role does not pick out a unique causal relation; it picks out miscellaneous, disjunctive facts; (b) the role does not appear to pick out anything in cases where omissions and absences are allowed as causes (cf. Lewis 2004).

This dialectic situation amounts to a challenge for interventionists: they have to provide a positive argument for the claim that the conceptual role of causation indeed picks out a role-filler. If such an argument can be established, interventionists could choose the Canberra plan as an option in order to adopt the SM-account. There is reason to be optimistic about the prospects of a successful defence of the Canberra plan (cf. Liebesman 2011, who addresses the two problems for the Canberra plan applied to causation mentioned in the previous paragraph). It seems to me a stimulating question for future research whether this is so. In defence of interventionists, one can maintain that whether the Canberra plan can be defended is still an open research question (for a detailed discussion of the prospects of the Canberra plan and of alternative strategies to deal with Streven's objection see Reutlinger 2013).

## Conclusion

This chapter began with an ontological problem for non-reductive theories of causation: conceptually non-reductive approaches to causation suggest the metaphysical claim that causation is ontologically fundamental and irreducible. This metaphysical claim is problematic because of various Russellian arguments for the claim that there is no causation in fundamental physics. Most importantly, Russell argues that causal relations are time-asymmetric, but the nomic relations in fundamental physics are time-symmetric. Russell concludes from this observation, together with the assumption that the fundamental laws are complete and global, that there cannot be any fundamental causation. Yet, Russell's claim is at odds with the practice of the special sciences: researchers in the special sciences seem to make causal claims and to test them. For this reason, contemporary neo-Russellians defend the claim that there are no causes according to fundamental physics (orthodox Russellian claim) and, still, there are higher-level causal facts of the special sciences (neo-Russellian claim). Interventionists advocate the open-systems argument for the neo-Russellian claim. I presented several objections against this argument. This is a bad result for advocates of an interventionist theory of causation for two reasons: first, the open-systems argument was supposed to be the positive argument for the indispensable theoretical value of using the notion of an intervention. Since the open-systems argument is unsound, we still have no positive reason to use the notion of a possible intervention in order to state truth conditions for causal claims. Second, if the open-systems argument fails, it is unclear how interventionists can meet the neo-Russellian challenge. The chapter continued by presenting the SM-account supporting the neo-Russellian claim as an alternative to the open-systems argument. Responding to a worry by Michael Strevens, it was argued that, if one can separate analysing concepts from doing metaphysics to a satisfying degree, there is no reason why the SM-account is not an option for non-reductive theories. The Canberra plan might offer a way for interventionists to separate conceptual and metaphysical issues. If this is correct, then proponents of a non-reductive theory of causation are not committed to take causation as ontologically fundamental. Thus, the alleged ontological problem of a commitment to fundamental causation for interventionists need not arise. For this reason, I believe that the SM-account is an attractive way of explaining why causation possesses the features of time-asymmetry and causal asymmetry within the framework of my comparative variability theory of causation (see Chapter 8).

# Part III
# An Alternative Theory of
# Causation in Special Sciences

# 7

# In Defence of Conceptually Non-Reductive Explications of Causation

## Conceptually non-reductive explications of causation

Chapters 4–6 attempted to achieve the negative goal of this book: they have attempted to establish a critique of Woodward's interventionist theory of causation. Chapter 4 argued for the claim that the notion of a possible intervention is highly problematic, because it is either dispensable or it leads to an inadequate evaluation of interventionist counterfactuals. It was concluded that this is a discouraging result for interventionists. Chapter 5 raised a follow-up problem resulting from the arguments in Chapter 4: interventionists, such as Woodward, use the notion of a possible intervention to analyse what it is to be a special science law (the interventionist invariance account of laws). Yet, if the arguments presented in Chapter 4 are sound, then one cannot maintain the invariance account of laws as it stands, because its success depends on the problematic notion of an intervention. Facing this problem, an alternative view of non-universal laws (i.e. 'lawish statements') in the special sciences was provided. My account of non-universal laws crucially relies on quasi-Newtonian laws and a refined notion of invariance* (that does not depend on interventions). The primary target in Chapter 6 was the open-systems argument that is entertained by several interventionists. The open-systems argument is supposed to explain (a) why causation in the special sciences has certain typical features (such as causal asymmetry, time-asymmetry and so on, as the naturalist criterion of adequacy requires); and (b) why the neo-Russellian claim is true (i.e. the claim that there are non-fundamental causal facts of the special sciences, but there are no causal facts at the fundamental physical level). The open-systems argument crucially relies on interventions. Several objections were provided regarding the open-systems argument and the conclusion was drawn that the open-systems argument is not sound. Therefore, the argument cannot perform the explanatory tasks that interventionists expect of it. The upshot is that interventionists fail to provide a positive argument for the (indispensable) theoretical value of interventions.

The focus of the book will now turn to the positive target (Part III). The interventionist theory of causation will now be replaced with my own account of causation in the special sciences: the comparative variability theory of causation. Chapter 7 will vindicate conceptually non-reductive explications of causation. Concerning the legitimacy of conceptually non-reductive theories of causation, I side with interventionists in defence of theories of this kind – most importantly for the reason that my own account shares the feature of non-reductivity with Woodward's interventionist theory. Chapter 8 will present and argue in favour of my comparative variability theory of causation. Concluding the book, Chapter 9 will describe the benefits of my theory of causation for other debates in philosophy of science: I will focus on the impact of the comparative variability theory on scientific explanation, representations of mechanisms, and the conditional analysis of dispositions.

In the recent debate on causation, several attempts of conceptually non-reductive explications of causation have been proposed. These explications are non-reductive, in the sense that a causal notion is explicated by means of (other) causal notions. There are prominent examples of the conceptually non-reductive approach, now presented in chronological order:

- Cartwright's classic paper 'Causal Laws and Effective Strategies' advocates a probabilistic explication of causation that is non-reductive because it refers to 'causally homogeneous situations' (Cartwright 1983: 25.
- As pointed out in Chapter 2, Woodward's (2003: 59) interventionist theory of causation is conceptually non-reductive because it refers to interventions (which are explicitly introduced as causal notions) and it includes a clause to hold other causes fixed.[1]
- Schaffer (2004b: 305) argues for a Lewisian counterfactual theory of causation. Yet, Schaffer's preferred semantics for counterfactuals is non-reductive because it employs a similarity measure that essentially depends on causal independence conditions.
- My own theory of causation, the comparative variability theory of causation (see Chapter 8), also exhibits this non-reductive feature,

because it explicates 'X causes Y' in terms of counterfactual dependence of Y upon X in situations in which other causes of the target causal relata X and Y are held fixed (I called these situations 'comparative ceteris paribus situations': see p. 142).

- It seems to me that even regularity views of causation (Mackie 1980; Psillos 2009; Strevens 2009) are in a similar situation of appealing to other causal notions – despite their prima facie appearance of being conceptually reductive – because these accounts presuppose a causal field or a causal background conditional on which a particular regularity is assumed to exist. The general upshot of these observations is that if the conceptually non-reductive character of a theory of causation is problematic, then this problem does affect far more theories of causation than Woodward's interventionist account.

What are the reasons to accept conceptually non-reductive theories of causation? In my opinion, there are two compelling reasons: first, they cope with alleged counter-examples extremely well (such as various scenarios of pre-emption and over-determination; more generally speaking, these theories conform to the distinction criterion, see Chapter 1 (pp. 13–15), and Chapter 2 (pp. 57–70) in which the success of interventionist theories is presented). A second reason to accept conceptually non-reductive theories of causation is that they seem to reconstruct examples from the special sciences adequately (i.e. these theories conform to the naturalist criterion by correctly reconstructing the kinds and features of causal relations in the special sciences: again, see Chapter 1 (pp. 15–21) and Chapter 2 (pp. 52–7)). Yet, despite these reasons in favour of conceptually non-reductive explications, one might still remain sceptical. One may wonder whether it is not the case that conceptual circularity is problematic because being circular prevents an explication from being informative in any way. In other words, the concern is that conceptually non-reductive explications of causation are trivial and, therefore, useless – as, for example, Psillos (2002, 2007) and Strevens (2007) argue. According to the critics, non-reductive approaches are (at least, at first glance) clearly at odds with traditional, empiricist methods of explication in philosophy of science. According to these methods, analysing the concept of causation consists in providing necessary and sufficient conditions for causal concepts in acausal terms; that is, by using a vocabulary that is free of any causal notions (cf. Carnap 1966: 189). This is the project of reductive explication. The intuitive appeal of reductive explication is that an analysis has to be reductive in order to be illuminating and to improve our understanding of a concept that is in need of clarification. And, most certainly, the threat of triviality is a serious challenge to the approaches that have been mentioned here. This chapter will attempt to defend the advantages of a non-reductive approach to causation against the concerns that are motivated by the traditional view.

The next section will briefly remind the reader of David Lewis's counterfactual theory of causation. Lewis's theory is paradigmatic in two ways: first, the counterfactual theory of causation is the current orthodoxy in the debate on causation and the received view of causation in other areas of philosophy (e.g. in the philosophy of mind). Second, the counterfactual theory of causation is also a paradigmatic example of a reductive analysis of the concept of causation. The chapter goes on to provide three examples of an analysis of causation that is non-reductive because these approaches use causal notions in the analysans: these are the analyses of causation of Nancy Cartwright, James Woodward and Jonathan Schaffer. Further, these non-reductive approaches are widely believed to satisfy two criteria of an adequate conceptual analysis of causation successfully (i.e. the naturalist criterion and the distinction criterion). Most importantly, these non-reductive analyses seem to satisfy the standards of adequacy even better than Lewis's reductive theory. The trade-off argument for non-reductive theories states that we give up the requirement of reductivity as a trade-off for successfully meeting the naturalist criterion and the distinction criterion. The result of this trade-off is to accept non-reductive theories. But can that be right? Is it not obvious to object to any non-reductive analysis that it is circular? The subsequent section will consider the objection that any non-reductive definition of causation is (in two senses) trivial because it is viciously circular and, thus, it is of no use. This objection is then countered and it will be argued that we can happily accept non-reductive theories of causation. Finally, five reasons are offered in defence of a non-reductive analysis.

## A reductive analysis of causation: Lewis's counterfactual theory

One of the currently most influential analyses of causation is David Lewis's (1973b, 2004) counterfactual theory of causation. Let me briefly recapitulate the reconstruction of Lewis's theory that was presented in Chapter 2 (pp. 57–9). Lewis analyses causation in terms of counterfactual dependence between propositions that state the occurrence of distinct events. More precisely, Lewis defines actual causation in two steps: first, he defines causal dependence; then, he defines causation based on causal dependence. According to Lewis (1973b: 165–7), for two distinct possible events $c$ and $e$, *e causally depends* on $c$ iff the following counterfactuals are true: (1) if $c$ were to occur, then $e$ would occur, and (2) if $c$ were not to occur, then $e$ would not occur. Lewis defines causation as a transitive relation: be $c$, $d$, $e$ ... a finite sequence of distinct events such that $d$ causally depends on $c$, $e$ causally depends on $d$, and so on. Lewis calls this sequence a causal chain. Lewis defines actual causation by using the notion of a causal chain: 'one event is the cause of another iff there exists a causal chain leading from the first to the second' (Lewis 1973b: 167). Lewis's counterfactual theory of causation is reductive because it makes no use of causal terms. Merely the *acausal* notion of counterfactual dependence is used. Suppose one might object thus:

> Well, the definition of causation *itself* does not use causal notions. But what about the truth conditions for *counterfactuals*? Is it not the case that one needs causal terms or, at least, causal intuitions in order to evaluate counterfactuals?

According to Lewis, this is not the case. Lewis's account of truth conditions for counterfactuals meets the conditions of a reductive analysis as well as his theory of causation. Recall Lewis's semantics for counterfactuals, presented in Chapter 3 (pp. 79–82):

> Lewisian semantics: The counterfactual *if $\phi$ were the case, then $\psi$ would be the case* is (non-vacuously) true at a possible world w if and only if $\psi$ is true in the closest $\phi$-worlds. (cf. Lewis 1973a: 16, 1973b: 164)[2]

The closeness of worlds is measured by a similarity heuristic. The criteria of similarity between worlds consist in shared facts about laws of nature and the amount of shared particular non-nomic facts (see Chapter 3, pp. 80–1, for details). Terms referring to particular facts and facts about laws of nature are part of the vocabulary that is approved by proponents of reductive analysis. Therefore, Lewis's analysis of causation, including his semantics for counterfactual conditionals, obeys the rules of a reductive methodology.[3]

## The trade-off argument for non-reductive approaches

In the debate on causation, reductive approaches have imposed several conditions of adequacy of analysis. The adequacy of an analysis is judged with respect to (at least) three criteria. Two of these criteria have already been introduced.

*Naturalist criterion*: According to this criterion, one asks: Does the analysis apply to paradigmatic causal claims in the sciences? If one intends to satisfy the naturalist criterion, then any theory of causation in the special sciences has to account for various kinds of causation and the features typically ascribed to causal relations.

*Distinction criterion*: According to this second criterion, one asks: Does the analysis distinguish correctly between intuitively different causal structures? For instance, the analysis is required to distinguish and describe correctly cases of over-determination, pre-emption, common cause scenarios and so on.

A third criterion that is central for the topic of the present chapter is often implicitly assumed to be a norm for conceptual analysis:

*Reductivity criterion*: The analysis of a causal notion should be reductive. That is, the analysans should only contain acausal notions. This criterion is supposed to guarantee that we achieve a learning effect when we analyse the notion of causation.

Although Lewis conforms to the third criterion, the first criterion and the second criterion have caused severe problems for Lewis's counterfactual theory of causation. For example, if we focus on the naturalist criterion, Lewis's theory only accounts for actual causation and not for various kinds of type-level causation. Also questions such as 'What corresponds to the similarity relation in the special sciences?' cast doubt on whether Lewis's approach is an adequate explication of causation in the special sciences. Concerning the distinction criterion, especially scenarios of pre-emption and over-determination have proven to be tenacious problems for Lewis's theory.[4] Of course, the failure to meet these criteria in some cases is not a knock-down argument against the counterfactual theory. The counterfactual theory might be improved by modifications in order to deal with counter-examples.[5]

However, independently of how the prospects of Lewis's counterfactual theory are, there is a problem: some theories of causation in the current debate appear to be more successful in meeting two criteria of adequacy of an adequate theory of causation in the special sciences (i.e. the Naturalist and the Distinction criterion) than Lewis's reductive analysis of causation (and other reductive theories of causation, such as probabilistic and regularity theories). It is a characteristic of these successful theories that they are non-reductive, because they make use of causal notions in the analysans.

This observation motivates the main argument for adopting non-reductive analyses: reductive theories of causation, such as Lewis's paradigmatic approach, fail to meet the naturalist criterion and the distinction criterion, although they satisfy the reductivity criterion. Even if one would prefer reductive theories of causation, one has to make the following observation: non-reductive theories, such as the interventionist theory, conform to the naturalist criterion and the distinction criterion in a superior way. In other words, non-reductive approaches seem to be better candidates for an adequate explication of causation in the special sciences. The ultimate reason to adopt non-reductive theories of causation is a trade-off argument: we choose non-reductive approaches because they have proven to be superior in satisfying the naturalist criterion and the distinction criterion. Satisfying these criteria outweighs the fact that non-reductive theories violate the reductivity criterion. Therefore, non-reductive theories (which are successful with respect to the naturalist criterion and the distinction criterion) are outstandingly promising candidates for an adequate explication of causation in the special sciences.

There are at least two reasons to believe that satisfying the naturalist criterion and distinction criterion outweighs a violation of the reductivity criterion. First, one might simply compare the number of criteria that a non-reductive theory satisfies to the number of criteria that a reductive theory satisfies. If this is the measure of success, then non-reductive theories (take the interventionist approach) satisfy a larger number of sub-criteria of the naturalist criterion and distinction criterion than a paradigmatic reductive theory (say, Lewis's counterfactual theory).

Second, one might want to assign different weights to the three criteria. I think there is a good reason to rank them in the following order:

- Naturalist criterion,
- Distinction criterion,
- Reductivity criterion.

If we accept this ranking, then non-reductive theories (e.g., Woodward's account) satisfy more criteria of greater importance than a paradigmatic reductive theory (e.g., Lewis's account). Hence, one ought to prefer the non-reductive theories of causation. The main reason to rank the criteria in this order is that scientific representations and practices should guide our commitment to various kinds and features of causation (at least, if the goal is to provide an explication of causal concepts in a scientific context). The causal scenarios considered by the distinction criterion provide us with information about whether and how we are able to apply causal notions used in the sciences (i.e. concepts grasping the kinds of causation with features of causation as revealed by the naturalist criterion) to the actual and possible causal situations. The reductivity criterion differs from the other criteria because it is not directly postulated for improving our understanding of the sciences. It is, rather, justified by the idea that we cannot learn anything about the concept of causation unless we analyse it in acausal and more fundamental terms. This latter claim is, however, controversial. It remains to be seen whether a better understanding of causation is necessarily tied to proceeding in a reductive manner. I deny precisely this claim undertaken by advocates of the reductive stance.[6]

Let me illustrate the non-reductive analysis of causation by three widely-acknowledged examples: the accounts of Nancy Cartwright, James Woodward, and Jonathan Schaffer. I will present these accounts in historical order (although Woodward's interventionist theory is of central importance).

### Example 1: Nancy Cartwright's theory of causation

In *Causal Laws and Effective Strategies*, Cartwright analyses causation (on the type level):[7] '*C* causes *E* if and only if *C* increases the probability of *E* in every situation which is otherwise *causally homogeneous* with respect to *E*.' (Cartwright 1983: 25) A situation is 'causally homogeneous with respect to *E*' if other causes $\{U_i\}$ of *E* are (experimentally or hypothetically) held fixed such that the other causes $\{U_i\}$ of *E* are not statistically relevant for *C*. Cartwright motivates the need for information about the other causes of E by the following kind of example. Suppose that smoking (*C*) causes lung cancer (*E*). Suppose further that physical exercising (*U*) prevents lung cancer. If one does not hold *U* constant (e.g. by examining only populations of smokers who exercise), then it could be the case that smoking does not increase the probability of lung cancer *although* smoking does cause lung cancer. This situation might occur if it is the case that Prob(no lung cancer | exercising) > Prob(lung cancer | smoking). Consequently, a situation with such a probability distribution is a counter-example to a naive probabilistic theory of causation which analysis causes as mere probability raisers for the occurrence of the effect. Cartwright concludes that one cannot analyse the concept of causation without referring to information about other causes of the effect *E* (Cartwright 1983: 23f).[8] For the reason that Cartwright's theory of causation essentially refers to other causes of the effect $E$, her analysis is clearly non-reductive.

## Example 2: Woodward's interventionist theory of causation

Interventionists such as Woodward advocate a modified counterfactual theory of causation, because their definitions of causation depend essentially on interventionist counterfactual conditionals. In this respect, the interventionist project resembles Lewis's counterfactual theory of causation. Interventionist counterfactuals are of the following form: If there were an intervention $I = i_i$ on $X$ such that the value of $X$ were changed to be $x_i$, then $Y = y_i$ would be the case. I have provided alternative ways to understand the truth conditions of interventionist counterfactuals. Despite the common feature of referring to counterfactuals, there is a striking difference between Lewisian counterfactual theories and interventionist counterfactual theories of causation. Although the interventionist theory of causation can be interpreted as a counterfactual approach to causation, the interventionist definition differs from the orthodox Lewisian counterfactual theory of causation in the methodological constraints on conceptual analysis. Woodward's interventionist theory of causation is *not reductive* because it involves two kinds of causal concepts: first, the concept of a possible intervention; and, second, in the same way as Cartwright, information about other causes of the effect $Y$ ('when all other variables are held fixed at some value by intervention', Woodward 2003: 55). Any definition with such features is clearly at odds with the reductive methodology. The first kind of causal notion – that is, the notion of an intervention – is a genuine feature of Woodward's approach (see Chapter 2: pp. 29 and 32). As already mentioned, Woodward's idea of information about other causes is close to, if not equivalent to, Cartwright's 'causally homogeneous situation'.

Obviously, Woodward's definition is non-reductive. Woodward describes his own project as clarifying the meaning of causal notions (cf. Woodward 2003: 7–9, 38; 2008: 195f). As I have indicated earlier, I agree with Michael Strevens that the best interpretation of Woodward's semantic project is to understand it as providing the truth conditions of causal statements (cf. Strevens 2008: 184). Lewis and Woodward agree on truth conditional semantics in order to determine the meaning of a causal statement. However, they disagree on the question of whether these truth conditions can be provided in causal terms.

**Example 3: Schaffer's non-reductive counterfactual theory**

Schaffer analyses causation in terms of counterfactual dependence (in agreement with Lewis). But Schaffer's approach employs a different similarity relation in the semantics for counterfactuals than Lewis:

- It is of the first importance to avoid big miracles.
- It is of the second importance to maximize the region of perfect match, *from those regions causally independent of whether or not the antecedent obtains.*
- It is of the third importance to avoid small miracles.
- It is of the fourth importance to maximize the spatiotemporal region of approximate match, *from those regions causally independent of whether or not the antecedent obtains* (Schaffer 2004b: 305, original emphasis).

Schaffer's similarity heuristic counts only those world-'regions' that are not (direct or indirect) effects of the antecedent. Although the analysis of causation is reductive, the semantics for counterfactuals clearly violates the reductivity criterion, because it appeals to causal independence conditions. Thus, Schaffer's theory of causation is clearly non-reductive. Schaffer himself acknowledges this feature of his analysis (cf. Schaffer 2004b: 307f.). However, his main reason for arguing for a non-reductive counterfactual theory of causation is also a trade-off argument: Schaffer abandons the reductive character of Lewis's theory of causation in order to deal with various counter-examples to Lewis's original theory. The fact that the new non-reductive version of the counterfactual theory is significantly superior to Lewis's original approach outweighs a violation of the reductivity criterion (cf. Schaffer 2004b: 299, 305–7).

Cartwright, Woodward, and Schaffer are not alone in advocating that non-reductive analysis is a legitimate and fruitful enterprise: apart from the tradition of providing necessary and sufficient truth conditions in a reductive way, there is also a longstanding non-reductive tradition in favour of views such as those of Cartwright, Woodward, and Schaffer (and also with respect to other topics in philosophy besides causation). Along the lines of this tradition, an informative conceptual analysis relates important concepts – such as 'intervention' and 'direct cause' – without pointing out a class of more fundamental concepts. In the literature, one finds many philosophers sympathetic to such a methodology. Famously, Peter Strawson – one of the most prominent adherents of the non-reductive tradition – distinguishes a reductive way to analyse concepts from a 'connective' – that is, non-reductive – way. According to Strawson, 'connective analysis' or 'connective elucidation' analyses the meaning of a concept 'only by grasping its connection with others' (cf. Strawson 1992: 19). Note that Strawson does not deny that there are reductives analyses. His main point is to emphasize that the method of connective analyses often seems (a) to fit philosophical practice better, and (b) that connection is illuminating in many cases. (cf. Strawson 1992: ch. 2; also Glock 2003: 115, 244)

## The objection against non-reductive approaches

The objection that might arise against a non-reductive definition is obviously a worry concerning circularity:

> A non-reductive definition of causation is trivial because it is viciously circular and, thus, it is of no use and it does not even amount to a proper analysis of causation. (cf. Psillos 2002: 104f., 182f.; 2007: 99; Strevens 2007: 245)

This is a serious objection that has to be refuted in order to maintain theories of causation in the style of Cartwright, Woodward, and Schaffer. But is this objection justified? Are conceptually non-reductive theories of causation really worthless because they use causal notions? Could we not accept a non-reductive definition because it tells us something interesting? One reason to accept these theories is the trade-off argument. However, the challenge for advocates of a non-reductive account is to show that their account is not viciously circular. In order to be clear on the concept of circularity involved in non-reductive theories of causation, let us distinguish between two senses of 'circular':

> **Circularity I:** The analysandum – say, '$X$ directly causes $Y$' – occurs in the analysans itself.

Following Boghossian (2000: 245), one might call this first kind of circularity 'gross circularity'. Boghossian connects two problems with gross circularity: First, gross circularity is question-begging, because 'this manoeuvre offends the idea of proving something or arguing for it'. Second, gross circularity faces the problem of bad company, because 'a grossly circular argument is able to prove anything, however intuitively unjustifiable'. As will be argued in accordance with Boghossian (and Woodward), grossly circular definitions of causation fail to be in the least illuminating.

The first kind of circularity can be distinguished from a different kind of circularity (cf. Burgess 2008: 219f.; Humberstone 1997: 250, who call it 'analytic circularity').

> **Circularity II**: A specific causal notion such as 'directly causes' (the analysandum) also occurs in the analysis itself. Yet, the relation of direct causation holds with respect to different entities in the analysans than in the analysandum.

Bearing these different meanings of circularity in mind, let us now explore five strategies to defend the trade-off argument for conceptually non-reductive theories of causation against various objections.

## Defending non-reductive explications

### Defence 1

Woodward defends the non-reductive analysis of causation with the following argument: the analysis is *non-reductive*, but it is – contrary to the objection raised on p. 187 – not viciously circular or a petitio principii in the sense of Circularity I. Let us consider what Woodward means by distinguishing being non-reductive and being viciously circular. Woodward insists that circularity is nothing bad at all, if it only leads to admitting to the fact that we are not capable of providing a completely reductive analysis of causation (cf. Hitchcock and Woodward 2003: 197). A reductive analysis deletes all causal notions from the analysans. But also, a non-reductive analysis may reveal interesting conceptual connections between the notions of causation, intervention, law, explanation, stability of counterfactuals and so on (cf. Woodward 2003: 103f.). The crucial point is that we can be non-reductive without being viciously circular; that is, circular in the sense of Circularity I. For instance, Woodward's definition of direct causation says, roughly, that $X$ causes $Y$ iff $Y$ is counterfactually dependent on $X$ when we change the value of $X$ by an intervention and all other variables are held fixed. We may willingly admit that 'to intervene' is itself a causal notion. But it is by no means true that we have to presuppose that $X$ causes Y in order to spell out what is meant by '$X$ causes $\underline{Y}$'. Doing so would, indeed, be viciously circular.

In a discussion with Woodward, Michael Strevens challenges Woodward's strategy by claiming that interventionist definitions are, indeed, viciously circular. Strevens considers the statement '$X$ is a direct type-level cause of $Y$' relative to a set V of three variables $\{X, Y, Z\}$. He implicitly assumes that we also know a set L of laws describing the relations among the variables in **V**. That is, we are considering a specific causal model ⟨**V**, **L**⟩ according to which '*X* is a direct type-level cause of *Y*' is assigned a truth-value. We assume that *Z* is another cause of *Y* besides *X*, and we also assume that *Z* lies on a path to *Y* that does not lead through *X*. Strevens argues that even in such a simple scenario interventionist notions of causation prove to be viciously circular (in the sense of Circularity I).

> I need an intervention on [*X*]. This intervention (itself represented by a variable [*I*]) must not be causally linked to any variable that is in turn a cause of [*Y*], unless the link goes by way of [*X*]. In particular, either it must not be a direct cause of [*Z*], or [*Z*] must not be a direct cause of [*Y*], in each case relative to [a set of variables **V**]. To rule out the first possibility, I have to find an intervention on the intervention variable [*I*]. Infinite regress, you are thinking. Actually, worse: even if the gods were to hand me such a variable, I am now testing for causation relative to a variable set that includes not only the original three variables but [*I*] as well. What I needed, however, were the facts about causation relative to just the three variables. Dead end. So perhaps I can show that [*Z*] is not a direct cause of [*Y*]? *To do this I need, by parity of reasoning, to settle the question of whether [X] directly causes [Y]. Circularity.* (Strevens 2007: 245, emphasis added)

Strevens's strategy seems to consist in pointing out the following two issues:

First, '*X* is a direct type-level cause of *Y*' is true relative to the causal model iff there is a logically possible intervention on *X* such that *Y* changes while *Z* is held fixed (i.e. *Z* is held fixed at a value in the redundancy range, one might add on Woodward's behalf). Strevens observes correctly that according to Woodward's definition of an intervention variable (2003: 98), among other conditions, *Z* must not be correlated with the intervention variable *I*, and *I* is required to be an indirect cause of *Y* via a path that leads through *X* (if *I* is a cause of *Y* at all). But what is problematic about this? I agree with Strevens that, once one asks what the truth conditions for '*I* is an indirect cause of *Y* via *X*' are, one has to introduce an intervention variable *I*\* for *I*. And, if one wants to know whether '*I*\* is an intervention variable for *I*' is true, then a third intervention variable *I*\*\* has to be introduced, and so on ad infinitum. Yet, given that we are interested in the truth conditions of the causal claim '*X* is a direct type-level cause of *Y*' relative to the original causal model, then the concern does not arise. From an interventionist point of view, introducing some intervention variable $I$ provides all the truth conditions for a causal claim relative to the causal model in question. And, even if there were the problem of an infinite regress, it would not amount to the objection that interventionist definitions are viciously circular (i.e. in the sense of Circularity I). The interventionist may interpret an infinite regress as an indication for the idea that causal concepts are fundamental concepts that cannot be reductively analysed.

Second, Strevens argues that the problem of a vicious circle arises, because knowing the truth conditions of '$X$ is a direct type-level cause of $Y$' relative to the causal model requires knowing whether $Z$ is a (direct type-level) cause of $Y$. Strevens is correct about the fact that an interventionist needs to know whether and how $Z$ influences $Y$ causally, because '$X$ is a direct type-level cause of $Y$' is true iff there is a logically possible intervention on $X$ such that $Y$ changes, given that $Z$ takes a value in the redundancy range. In turn, '$Z$ is a direct type-level cause of $Y$' is true relative to the same model iff there is a logically possible intervention on $Z$ relative to $Y$ such that $Y$ changes, given that $X$ takes a value in the redundancy range. I wonder where exactly the vicious circularity (i.e. Circularity I) is supposed to enter. I agree with Strevens that the interventionist starts with a causal model that contains all the causal information. According to the laws, we know all the relations between the variables in the model. We use these laws to evaluate interventionist counterfactuals, as required by the definition of a direct type-level cause. Is this a case of vicious circularity as Strevens suggests? I do not think this is the case. The case that Strevens presents simply shows that the truth conditions of '$Z$ is a direct type-level cause of $Y$' include different causal background knowledge than '$X$ is a direct type-level cause of $Y$'. In order to apply the interventionist definition of direct type-level causation, the former claim requires knowing whether $X$ is an off-path cause of $Y$. The latter claim requires knowing whether $Z$ is an off-path cause of $Y$. Does this teach us that interventionist definitions are viciously circular? Not at all. Strevens's example teaches us to acknowledge that evaluating different causal claims relative to the same causal model often requires different causal background assumptions (cf. Woodward 2008: 204).

Cartwright, Woodward, and Schaffer are likely to agree with philosophers such as Strevens who accept the reductive methodology that any alleged analysis that is viciously circular (in the sense of Circularity I) is automatically ruled out as an analysis. A *viciously circular* analysis is not tested whether and how it meets the criteria of adequacy for an analysis. *It is no analysis at all.* However, it has to be emphasized that the range of Woodward's reaction to the (ambiguous) charge of circularity is limited to Circularity I. What remains to be shown is why an analysis of causation can be Circular II without being vicious or useless.

## Defence 2

The concerns in respect of non-reductive analysis often seem to stem from the idea that analysing a concept is nothing but providing an explicit definition. The main purpose of explicit definitions consists in providing a synonym (a definiens) for the term to be defined (the definiendum) (cf. Suppes 1957: ch. 8); for example, a definition of mathematical operators such subtraction or division. Paradigmatically, Suppes (1957: 154) introduces two criteria that constrain the candidates of possible definitions: eliminability and non-creativity. According to the criterion of eliminability, the definiendum can be replaced salva veritate by the definiens. According to the criterion of non-creativity, one should not be able to derive any new conclusions from the (language containing the definiendum and the) definiens that cannot be derived from (a language containing only) the definiendum.

Obviously, using the definiendum in the definiens undermines the task of providing a synonym that is helpful in any way. This would be a problem for any philosopher trying to analyse causation in a non-reductive way, if 'analysing' causation would literally mean to provide an explicit definition of causal notions. But this does not seem to be the case. Rather, analysing causation in philosophy of science is better reconstructed as an explication in the sense introduced by Rudolf Carnap. In his *Logical Foundations of Probability*, Carnap characterizes an explication as follows: 'The task of an explication consists in transforming a given more or less inexact concept into an exact one or, rather, in replacing the first for the second' (Carnap 1950: 3, original emphasis). In philosophy of science, concepts such as causation, explanation and law have been regarded as 'more or less inexact'. Hence, philosophers of science have attempted to transform a given more or less inexact concept into one that is exact. However, as we shall see shortly, replacing an inexact concept with an exact one does not require a reductive, explicit definition. Precision can also be gain by an implicit or connective definition.

Although Carnap sometimes claims that an explication might have the form of an explicit definition, he carefully points out that, strictly speaking, there is an important difference between explications and explicit definitions:

I think that the analogy with the terms 'definiendum' and 'definiens' would not be useful because, if the explication consists in giving an explicit definition, then both definiens and definiendum in the definition express the explicatum, while the explicandum does not occur. (Carnap 1950: 3)

So, according to Carnap, the following problem arises: if one introduced an equivalent expression (the definiens) for the definiendum, then one would not have made the definiendum more precise as required by an explication. In this case, as Carnap says, 'the explicatum does not occur'. No progress has been made. In other words, if the explication were understood in terms of an explicit definition, then the task of providing a more precise concept (the explicatum) would not be carried out. As noted in Chapter 1, Carnap (1950: 7) believes that an explication (of causation) is adequate iff it conforms to the following criteria to a satisfying degree:

• The explicatum has to apply to paradigm cases of the explicandum.
• The explicatum has to be stated in an 'exact form'. It has to be stated by a 'well-connected system of scientific concepts' (which include concepts such as laws of nature, probabilities and so on).
• The explicatum has to be (potentially) fruitful for empirical scientific research.
• The explicatum should be as simple as possible; this means as simple as the more important requirements 1, 2, and 3 permit.

How does this relate to the criteria of adequacy introduced earlier? Let me briefly repeat the results of a more detailed discussion from Chapter 1.

Carnap's first criterion demands that the explicated concept (the explicatum) has to apply to paradigm cases of the concept that is supposed to be explicated (the explicandum). The naturalist criterion meets the first of Carnap's criteria, because (a) paradigm cases of causal statements, (b) paradigmatic features of causation, and (c) paradigmatic kinds of causation are provided by analysing case studies from successful sciences. Carnap's first criterion also directly relates to the distinction criterion of adequacy, because the explicated concept has to be able to distinguish correctly between intuitively different and intuitively possible causal situations (such as scenarios of pre-emption and common cause scenarios). The methodological justification of this criterion consists in the idea that the application of concepts ought to be tested for actual and merely possible causal situations.

Carnap's second criterion requires that the explicatum has to be stated in an 'exact form'. An explication is given in an exact form if it uses a 'well-connected system of scientific concepts'. The second of Carnap's criteria does not explicitly preclude that cause is a member of the family of scientific concepts (to which counterfactual or probabilistic dependence also belong). Note that the only constraint imposed on the notion of a cause by the second criterion is that it is made more precise (e.g. by formal means). In other words, a precise, explicated notion of cause may figure in the explicans.

Carnap's third criterion does not pose a problem for the non-reductive approach, because the explicated concept of causation is (a) the concept that is already 'fruitfully' (and often implicitly) used in the special sciences, and (b) a 'fruitful' tool for a (normative) methodology of discovering causes (although the extent to which Bayesian networks approaches to causal methodology can be defended is controversial – cf. Cartwright 2007; Glymour 2010).

Carnap's fourth criterion requires that the explication should avoid unnecessary assumptions in order to satisfy the other three criteria. This criterion is only a threat to non-reductive approaches if there are competing reductive accounts of causation that are superior – or, at least, equally good – in satisfying the other criteria (see Defence 3: pp. 194–5).

The decisive question is whether Carnap's method of explication implies the reductivity criterion. Although Carnap has occasionally argued in favour of the reductivity criterion, it seems that a non-reductive way to explicate a concept does not violate any of Carnap's criteria of an adequate explication. Both reductive and non-reductive approaches to the concept of causation seem to agree on the importance of Carnap's first criterion. Carnap's second criterion does not explicitly preclude that 'cause' is a member of the family of scientific concepts. Also, Carnap's third criterion allows non-reductively explicated concepts to be fruitful in the sciences (indeed, Bayesian networks methods might be a good example for the empirical success of a concept of causation that is not reductively defined). The lesson to learn is that Carnap's idea of an explication does not preclude that one can precisify a concept with a non-reductive explication.

However, Carnap even goes a step further. It is interesting that Carnap (1960: 228) explicitly allows explications to have the form of implicit or contextual definitions, or axiomatizations.[9] Because Carnap allows explications to have various logical forms (such as explicit and implicit definitions), he neutrally talks about 'meaning postulates' when he refers to the form of an explication (cf. Carnap 1960: app. B; 1966: chs 27 and 28). Further, a plausible way to understand a non-reductive analysis of causation that takes the form of an explication is in terms Strawson's connective analysis. Explications may be given in the form of a 'connective analysis' that analyses the meaning of a concept 'only by grasping its connection with others' (cf. Strawson 1992: 19; cf. also Skyrms 1980b, Burgess 2008: 231). For instance, Strawson takes semantic concepts, such as meaning and truth, to be connectively defined; that is, not reductively defined but 'only by grasping its connection' to other semantic concepts. The characterization of a connective analysis seems to be at least similar to an implicit definition of causation (cf. Boghossian 1996: 376; Hale and Wright 2000: 307–11). A contemporary ancestor of Strawson's connective analysis and implicit definitions is the idea of the theoretical role played by a concept as advocated by functionalists in the philosophy of mind and by philosophers engaged in the Canberra Plan, most importantly by Frank Jackson (1998). The theoretical role of a concept allows loosening the reductive methodology. Let me briefly sketch Jackson's idea of the theoretical role of concepts. According to Jackson (1998: ch. 2),[10] it is the goal of conceptual analysis to explicate the role of a concept in a (folk or scientific) theory. For instance, Jackson takes the notion of a fish to be analysed in the following way: 'x is a fish iff x has the important properties out of or descended from or explanatory of $F_1$, $F_2$, $F_3$, ..., according to the best true theory' (Jackson 1998: 35). Most importantly, Jackson claims that the reference of a concept is fixed by its theoretical role (as in the case of a connective analysis and as in an implicit definition) and the theoretical role of a concept is not necessarily determined in a reductive way. If we apply this idea to causal notions, then we may hold that the causal role might include folk features, some of which are characterized in a non-reductive way (see Chapter 6 (pp. 171–3), where I discuss the Canberra Plan as applied to causation from an interventionist perspective). Analogously to Jackson's idea, the non-reductive definition of direct causation is regarded best with respect to satisfying certain standards (e.g. the naturalist criterion and distinction criterion).

To sum up, the charge of circularity often seems to be motivated by the idea that analysing causal concepts is providing an explicit definition. At least for philosophy of science, many attempts to analyse the meaning of some concept differ from explicit definitions: instead of finding a synonym, the task usually is to make a vague concept (e.g. causation, explanation, law) more precise. A way to spell out this idea is Carnapian explication. Fortunately, the criteria of adequacy for an explication do not rule out a non-reductive approach.

## Defence 3

Suppose one accepts non-reductive conceptual analyses. Is it true that one has to give up the preference for reductive analysis? Certainly not.

In defence of non-reductive analysis, one merely has to claim that there are criteria of adequacy that a non-reductive analysis might meet in an overall better way than a reductive analysis (this is the trade-off argument developed earlier in this chapter). Those criteria consist in, first, accounting for kinds and features of causation as referred to in the special sciences (naturalist criterion); and, second, distinguishing intuitively different causal structures correctly (distinction criterion). Nevertheless, a proponent of non-reductive explication can allow for the preference for a reductive analysis in the following way: were these standards of adequacy equally met by a reductive as well as by a non-reductive analysis, one would choose the reductive analysis. Yet, such a situation is not actually case: one reason not to adopt Lewis's counter-factual theory is the fact that it struggles with the naturalist criterion (e.g. it does not account for various kinds of type-level causation) and the distinction criterion (especially with respect to various scenarios of pre-emption).[11] Thus, accepting non-reductivity can be due to a trade-off with respect to other criteria of adequacy (i.e. the distinction criterion and the naturalist criterion).

## Defence 4

Although one might not give up the preference for reductive explications (see Defence 3), one important positive reason for accepting a non-reductive analysis of causation is due to the standard of reconstruction of paradigmatic causal claims in scientific practice. Both Cartwright and Woodward define causation relative to models that represent the relevant causal factors for a phenomenon (e.g. lung cancer).[12] It is a characteristic of these models to include information about other relevant factors besides the causal factors $X$ and $Y$ in question. In other words, many typical causal claims in the special sciences are of the form '$X$ causes $Y$ – given that other causes of $Y$ are held fixed'. In other words, these causal claims openly refer to other causes. This characteristic feature of causal claims in the special sciences is mirrored by a sub-criterion of the naturalist criterion (the context-sub-criterion, see Chapter 1, p. 17). If one is a naturalist about causation, then examples of causal statements from the special sciences should guide our explication of causal notions. As opposed to such paradigmatic causal claims, our intuitions concerning causation have little influence on our choice of an explication of causation (or they are, at least, overridden by paradigmatic causal claims in the sciences). Thus, if we are naturalists, then we have a positive reason to accept non-reductive analyses (cf. Burgess 2008: 232, who argues in a similar way in order to defend a circular definition of colours).

Some philosophers will not be impressed by the observation that many typical causal claims in the (special) sciences are of the form '*X* causes *Y* – given that other causes of *Y* are held fixed'. They might insist that this observation does not change their goal to explicate causation in a reductive way. These philosophers are certainly entitled to believe that there is no knock-down argument against the conceptually reductive approach in general. However, an additional reason to believe that the truth conditions for a causal statement partly consist of a causal background is supported by Cartwright's argument for non-reductive explication. Cartwright argues that one cannot analyse the concept of causation without referring to information about other causes of the effect *E* (Cartwright 1983: 23f.), because otherwise – that is, if merely reliant on acausal correlations – 'any association [...] between two variables which holds in a given population can be reversed in the sub-populations by finding a third variable which is correlated with both' (Cartwright 1983: 24).

## Defence 5

Although an analysis of causation – taken just by itself – may be non-reductive, one may achieve something sufficiently similar to reduction by adding further assumptions from causal modelling. Judea Pearl (1988: ch. 3) and Christopher Hitchcock (2010: s. 3.7) argue for a criterion of uniqueness, which is supposed to be a close substitute for the reductivity criterion. According to the criterion of uniqueness, we expect an analysis of causation to assign a unique causal structure to some non-causal information (e.g. purely probabilistic information). Suppose there is a probability measure over a set of random variables. How can we assign a unique causal structure to this set? According to both Pearl and Hitchcock, we can do so by adding further assumptions to our analysis of causation: first, one adds temporal information to the variables. Second, one assumes that the probability distribution over the set of variables satisfies the causal Markov condition (which states that, conditional on the direct causes of *X*, the indirect causes of *X* are probabilistically irrelevant, see Chapter 2, p. 47). Although the causal Markov condition is also not a reductive (purely probabilistic) analysis of causation, it enables us to assign a unique causal structure to a given temporally ordered probability measure. What does this tell us? If an analysis of causal notions plus other assumptions (e.g. probability measure, temporal information, causal Markov condition) satisfies the criterion of uniqueness, then one might not have a straightforward reductive analysis. One may have a one-to-one mapping of a certain causal structure and a particular probability measure. As Hitchcock argues, any non-reductive explication of causation that meets the criterion of uniqueness is not hopelessly under-determined by empirical data (such as probabilistic correlations) as the critics of non-reductive approaches may suspect.

To sum up this section, I have presented five strategies in defence of the trade-off argument for non-reductive explications of causation. The most important result is that non-reductive explications can be defended against being viciously circular. According to the trade-off argument in favour of non-reductive explications of causation, one is willing to accept a non-reductive explication after the style of Cartwright, Woodward, and Schaffer depending on how well the non-reductive explication satisfies the naturalist criterion and distinction criterion of adequacy.

## Conclusion

I have presented several non-reductive explications of causation. These approaches are at odds with a traditional reductive constraint on conceptual analysis. I introduced the trade-off argument in favour of non-reductive theories of causation: using non-reductive approaches is justified because, in the debate on causation, they are taken to meet two criteria of adequacy in a better way than, for instance, the Lewisian reductive approach. The problem for non-reductive approaches is the objection of triviality. I presented five strategies in defence of non-reductive theories of causation in the special sciences:

1. They are not viciously circular.
2. They need not be understood as explicit definitions – rather, they are explications involving implicit definitions, axiomatizations, or connective analyses.
3. One need not give up a preference for reductivity. Accepting non-reductive analysis can be a matter of trade-off with respect to other criteria (i.e. the distinction criterion and the naturalist criterion).
4. Many paradigmatic causal statements in the special sciences refer openly to other causal factors. This feature of causation in the special sciences is mirrored by the context-sub-criterion of the naturalist criterion (see Chapter 1, p. 17). Thus, there might be a naturalistic reason to accept non-reductive explications.
5. A non-reductive explication of causation (plus additional assumptions) can satisfy the criterion of uniqueness; that is, there is a unique mapping of a causal structure to a particular probability distribution.

I conclude that non-reductive explications of causation can be vindicated. This result is important, because I will have to rely on it in Chapter 8 where I present my own explication of causation, the comparative variability theory of causation. Although my theory does not require the troubled notion of a possible intervention, it is conceptually non-reductive.

# 8

# The Comparative Variability
# Theory of Causation

## From interventions to comparative variability

So, where are we? Let me briefly summarize what I have presented and
argued for so far. Part I set the stage for the discussion about causation in the special sciences. Chapter 1 elaborated criteria of adequacy
for an explication of causation in the special sciences (the naturalist
and the distinction criterion). Chapter 2 introduced the influential
interventionist theory of causation. I focused on Woodward's widely-
received interventionist approach according to which $X$ causes $Y$ iff,
roughly, there is a possible intervention on $X$ that changes the value of $Y$.
I argued that interventionist theories are prima facie (promising to be)
in accord with the criteria of adequacy, although I diagnosed that there
are several desiderata that have to be addressed by interventionists (most
importantly, interventionists ought to account for the explication of a
non-universal law, and for the features of time-asymmetry and causal
asymmetry).

Part II was devoted to several objections to Woodward's interventionist theory. In Chapters 3–6, I pointed out serious and genuinely internal
problems for interventionist theories of causation, although the interventionist theory prima facie appeared to be adequate. Chapter 3 argued
that interventionists fail to specify the truth conditions of interventionist
counterfactuals. As an account of truth conditions is a desideratum,
I proposed three alternative semantics for interventionist counterfactuals (i.e. a possible worlds version and a Goodmanian version of
interventionist semantics, and an interventionist version of the suppositional theory that states assertability conditions). Chapter 4 presented
two arguments to the conclusion that the interventionists' key notion
of an intervention is problematic because of the modal character that

is attributed to interventions: Woodward requires that interventions be merely logically possible. This modal character of interventions leads to a dilemma for interventionists: either interventions are dispensable for the semantic project concerning causation, or interventions lead to an incorrect evaluation of interventionist counterfactuals. Chapter 5 addressed a follow-up problem for interventionists: if the notion of a possible intervention is troubled, then this also affects the account of laws that interventionists have proposed (the invariance theory of laws), because this theory crucially relies of the notion of an intervention. I continued to argue for the claim that one can account for the no-universal-laws requirement (i.e. one of the sub-criteria of the naturalist criterion) without appealing to the notion of an intervention by defending an alternative explication of (non-universal) laws in special sciences. Chapter 6 discussed the interventionist argument, the open-systems argument, that is supposed to explain the time-asymmetry of causation and the causal asymmetry. The open-systems argument essentially depends on the possibility of intervening. Because of the alleged explanatory power of the open-systems argument, interventionists believe that the argument provides a positive reason to use the notion of an intervention in order to explicate causation. However, I argued that the open-systems argument is not sound. If the open-systems argument is not sound, then we lack a positive and compelling reason to believe that interventions are indispensable for the enterprise of explicating causation. Moreover, interventionists fail to provide an explanation of the features of causal asymmetry and time-asymmetry. I have pointed out an alternative way to account for the time-asymmetry of causation: the statistical-mechanical approach by Albert (2000), Kutach (2007), and Loewer (2007, 2009). Chapter 6 was the only part of the book that explicitly addressed the metaphysics of causation and, thereby, deviated from the main semantic project of the explication of causal concepts.

Finally, Part III has the goal of developing an explication of causation in the special sciences that (a) is adequate, and (b) does not suffer from the severe internal problems of the interventionist theory. I will argue for what I call the *Comparative Variability theory of causation*. However, despite the objections to the interventionist theory, my own theory honours the achievements of Woodward's approach. The comparative variability theory preserves several features of the interventionist theories – and, in my opinion the partial success of the interventionist theory is due to these features, not due to the notion of an intervention. Chapter 7 addressed the issue that interventionist definitions of causal notions are conceptually circular. However, despite the criticisms of Chapters 3–6, I have defended the *conceptually non-reductive feature* of interventionist definitions. Being conceptually non-reductive is a feature that is not genuine to interventionist theories of causation. As Chapter 8 will show, my own theory of causation shares this feature with the interventionist theory – however, without relying on the notion of a possible intervention.

Given the arguments presented in Chapters 3–6, I am convinced that the original interventionist theory of causation can no longer be convincingly upheld. Especially, it no longer seems promising to rely on the notion of a possible intervention in order to master semantic and metaphysical tasks. A crucial question: Where shall we go from here? One available strategy could consist in turning away from the interventionist approach completely; for instance, by adopting a probabilistic theory of causation (such as Granger-causation in economics). Yet, it may be too hasty and premature to throw everything overboard, because the interventionist approach was quite successful in satisfying several sub-criteria of the naturalist criterion and the distinction criterion. For this reason, I will opt for a more preservative strategy: I will propose a departure from interventionist theories of causation that is supposed to be just as minimal so as to avoid its problems and to preserve its virtues. Following this strategy, I think that two essential features of interventionism can be vindicated: first, the claim that the truth conditions of causal statements can be stated in terms of non-backtracking counterfactual dependence among effect and cause. Second, the claim that counterfactual dependence between effect and cause is somehow conditional on holding fixed *other* causes of the putative effect. This strategy is a straightforward reaction to the dispensability argument and the argument against the incorrect evaluation of interventionist counterfactuals: if one is compelled to get rid of interventions because of these arguments, then the remaining features of the otherwise successful interventionist theory boil down to a counterfactual theory of causation that pays special attention to ceteris paribus clauses (i.e. to clauses that hold other causes fixed).

Historically, the idea that causal statements need to be qualified by a ceteris paribus clause, or by the assumption that other causes are held fixed, has been famously presented by the economist Alfred Marshall in the late nineteenth century. Marshall advocated and popularized the use of ceteris paribus clauses to flesh out the idea of holding other causes fixed. It was Marshall's genuine contribution to economics to advocate partial equilibrium analysis. Marshall claimed that an analysis of this kind holds merely 'ceteris paribus'. In his influential *Principles of Economics*, Marshall defines the task of economists in terms of the phrase 'ceteris paribus':

> [An economist answers] a complex question, studying one bit at a time, and at last combining his partial solutions into a more or less complete solution of the whole riddle. In breaking it up, he segregates those disturbing causes, whose wanderings happen to be inconvenient, for the time in a pound called Caeteris Paribus. (Marshall 1890: 366)

In other words, Marshall claims that economists typically suppose that, if *X* causes *Y*, other disturbing causes are held fixed (they are 'for the time in a pound called Caeteris Paribus'). In the introduction to the same work, Marshall explains why economics is interested in the isolation of causes by assuming that *other things are equal*:

> The forces to be dealt with (in economics) are, however, so numerous, that it is best to take a few at a time; and to work out a number of partial solutions as auxiliaries to our main study. Thus we begin by isolating the primary relations of supply, demand and price in regard to a particular commodity. We reduce to inaction all other forces by the phrase '*other things being equal*': We do not suppose that they are inert, but for the time we ignore their activity. This scientific device is a great deal older than science: it is the method by which, consciously or unconsciously, sensible men have dealt immemorial with every difficult problem of ordinary life. (Marshall 1890: xiii, emphasis added)

Marshall's key idea seems to be that, because of the messy complexity of their domain, economists are committed to causal statements of the form 'ceteris paribus, *X* causes *Y*', or, expressed in a language closer to interventionists, 'given other influences are held fixed, *X* causes *Y*'. As Marshall points out, the ceteris paribus clause is interpreted in the sense that other causes are *held constant* as opposed to the reading that other causes are *absent*: 'We do not suppose that they [that is, all other causal influences] are inert, but for the time we ignore their activity' (Marshall 1890: xiii). In the context of economics and the social sciences, it is seldom practically possible to hold a variable constant in an experimental set-up – instead, to hold causes constant amounts to the *theoretical* enterprise of postulating a model in which certain variables are held fixed.[1] This distinction of holding fixed and assuming the absence of certain variables is analogous to Schurz's distinction of the *exclusive* and the *comparative* reading of ceteris paribus clauses: the exclusive reading of the ceteris paribus clause expresses that other disturbing causes are *absent*, the comparative reading expresses that other causes are held fixed (cf. Schurz 2002; Reutlinger et al. 2011).

I will combine Marshall's idea of a ceteris paribus restriction of causal statements with a counterfactual theory of causation. This strategy aims to exploit the two essential features of interventionism that were already introduced above: first, the claim that the truth conditions of causal statements can be stated in terms of counterfactual dependence; and, second, the claim that counterfactual dependence is qualified by a comparative ceteris paribus clause – that is, counterfactual dependence is somehow conditional on holding fixed *other* causes of the putative effect. My allies in the present debate are the econometrician Heckman (2008), whose concept of causation is based on the notion of controlled variation (i.e. 'variation in treatment holding other factors constant'; Cartwright (2007), who advocates variation in 'epistemically convenient systems'; Menzies (2007), who explicates causation in terms of counterfactual dependence in a causal context; Field (2003), who explicates causation in terms of conditional counterfactual dependence (i.e. conditional on holding other factors fixed), and Hitchcock[2] (2001), who relies on the notion of an active causal path and the assumption to hold those causes fixed that are not on the directed causal path from the cause and the effect in question.[3]

Following Schurz's terminology, the *comparative* part of my theory of causation captures the ceteris paribus clause attached to causal statements in the special sciences. The *variability* part refers to the idea that I understand causation as difference-making, and, more particularly, as counterfactual dependence. Phrasing things a little carelessly, one could say that the comparative variability theory is interventionism without interventions.

In this chapter, I will propose that a comparative variability theory of causation can replace the original interventionist theory. In the next section, I will introduce a semantics for counterfactual conditionals (comparative variability semantics). The chapter goes on to provide explications of various causal notions which are based on the semantics introduced, and then shows that the comparative variability theory is in accord with the criteria of adequacy for an explication of causation in the special sciences. One upshot of the comparative variability theory is that it supports the dispensability argument against interventions because it does not appeal the possible interventions and it satisfies more criteria of adequacy than the interventionist account.

## Comparative variability semantics for counterfactuals

The core idea of the comparative variability theory of causation can be summarized as follows: first, the truth conditions of causal statements can be stated in terms of counterfactual dependence among effect and cause; and, second, counterfactual dependence is somehow conditional on the holding fixed of other causes of the putative effect. So, the theory has to answer two questions: first, what are the truth conditions of the counterfactuals that the definition of a causal notion uses?; and, second, to what does 'ceteris paribus' and 'holding fixed other causes of the effect' amount?

In order to answer these questions, let me briefly turn to the (deficient) interventionist approach to counterfactuals. Although this approach is deficient, I will use one of its central concepts: the notion of a redundancy range. The notion of a redundancy range is a handy formal recipe for how to hold other causes fixed.

As noted in Chapter 3, interventionists do not provide a detailed account of a semantics for counterfactuals. However, Woodward and Hitchcock (2003: 13f.) suggest that the counterfactual *if $X = x$ were the case, then $Y = y$ would be the case*[4] is true iff there is a logically possible world in which the following three conditions are all satisfied: (a) there is a possible intervention $I = i$ on $X$ relative to $Y$ such that $X = x$, (b) all other variables take their actual values, and (c) the consequent $Y = y$ is true. Conforming to interventionist definitions of causation,[5] one may improve and weaken condition (b) by merely requiring that all other variables take values that are in the so-called *redundancy range* (cf. Woodward 2003: 83; Hitchcock 2001: 290; see pp. 37–8) instead of requiring that they take their *actual* values. Let me briefly recapitulate what the redundancy range is because it is an important tool for my theory of causation. Suppose that '$X$ causes $Y$' is true. The redundancy range is a range of *actual and counterfactual* possible values that a variable might have, which lies off the directed path P (in question) from the cause $X$ to the effect $Y$. The redundancy range is defined with respect to other direct causes $W_1, ..., W_n$ of $Y$ – both (a) causes on a different path from $X$ to $Y$ and (b) causes of $Y$ that do not lie on any path that leads through $X$. Let me re-introduce the notion of a redundancy range for the actual values of variables on path P from $X$ to $Y$: the values $w_1, ..., w_n$ of $W_1, ..., W_n$ are in the redundancy range with respect to path P from $X$ to $Y$ iff it were the case that $W_1 = w_1, ..., W_n = w_n$, then the actual values of the variables on path P would not change. Although Woodward and Hitchcock define the redundancy range with respect to *actual* values of the cause $X$ and the effect $Y$, we can also define the redundancy range for non-actual possible values of $X$ and $Y$. Suppose, for simplicity's sake, that this path connects only two variables, $X$ and $Y$. The values $w_1, ..., w_n$ of $W_1, ..., W_n$ are in the redundancy range with respect to (merely possible values of the variables on) path P from $X$ to $Y$ iff there is a possible value $x_p$ of $X$ and there is a possible value $y_p$ of $Y$, and if it were the case that $W_1 = w_1, ..., W_n = w_n$, then it would be the case that $X = x_p$ and $Y = y_p$. This understanding of the redundancy range will be beneficial for the explication of various kinds of type-level causation. For illustrative purposes, one may say that to require that the values of other causes be in the redundancy range is a special case of Field's (2003: 452) conditional counterfactual dependence. A dependence relation of this kind is expressed by conditionals of the form 'if the value of $X$ had been $x^*$, and the values of variables that lie along other routes from $X$ to $Z$ were held fixed, then the value of $Z$ would have been different'.[6]

What kind of a semantics for counterfactuals does this suggest? Let us take a step back to recall which claims Woodward and other interventionists consider to be true about causal and counterfactual statements:

- causal statements and counterfactual claims have meaning;
- the meaning of causal and counterfactual statements is determined semantically by truth-conditions;
- in order to state truth-conditions one needs to invoke causal notions.

If we take these characterizations seriously, I argued in Chapter 3, there are two promising ways for interventionists to determine the truth conditions of counterfactuals: a modified possible worlds semantics and a modified Goodmanian semantics. A third alternative, the suppositional approach, is ruled out because it fails to conform to the three above-mentioned characterizations, because the suppositional approach abandons the view that meaning is fixed by truth conditions (rather, the meaning of a sentence consists of assertability conditions). Unfortunately, these two alternative semantics are only prima facie options for the interventionists. As I argued in Chapter 4, assuming logically possible interventions leads into a dilemma: either interventions being indispensable for the truth conditions of causal claims and of counterfactual conditionals is highly problematic; or interventions lead to the inadequate evaluation of counterfactuals. I argued that adopting an interventionist semantics does not avoid the dilemma. Thus, the conclusion of Chapter 4 was that we get rid of interventions.

Resulting from my objections to Woodward's interventionist theory of causation (see Chapters 3-6), I now propose to replace interventionist semantics with a close relative that is also conceptually non-reductive with respect to causal concepts and, yet, does *not* use the notion of an intervention. I will label this semantics *comparative variability semantics* (CVS). Analogously to interventionist semantics, CVS can be interpreted as a possible worlds semantics or as a Goodmanian semantics.

### The possible worlds version of CVS

A counterfactual *if $\phi$ were the case, then $\psi$ would be the case* is (non-vacuously) true at a small world w (relative to a causal model M = $\langle$U, V, L$\rangle$) if and only if $\psi$ is true in the closest $\phi$-worlds.

In analogy to the interventionist version of possible worlds semantics and in accord with Lewis's original intention, I maintain that the closeness relation in CVS is a *weak ordering relation*. That is, (a) a (small) world w might be equally close to two (small) worlds u and v, and (b) any two (small) worlds are comparable in their closeness. I propose to measure closeness and select the closest worlds with following the criteria:

### CVS selection function for the closest worlds

A world u is among the closest $\phi$-worlds accessible from w if the following three conditions are all satisfied in u:

- the statement $\phi$ (the antecedent) is true;
- all other direct causes $W_1$, ..., $W_n$ of the consequent (that are not on the considered path P from the antecedent variable to the consequent variable) take values that are in the redundancy range with respect to path P; and
- the set of laws L that holds in world w also holds in u; that is, the same functional relations obtain among the variables in w and u.

Worlds are more distant than the closest worlds to the degree that (a) they instantiate different laws, and (b) they contain variables that do not take values in the redundancy range. In analogy to Lewis's similarity heuristic, it is reasonable to assign a higher importance to differences in laws in order to measure the distance between small worlds. However, differences in 'non-nomic facts' – that is, the assignment of values to variables – are also important to measure the distance between worlds, because a world might be nomologically possible but all the relevant variables might fail to take values in the redundancy range.

Let me illustrate how the possible worlds version of CVS works by evaluating the counterfactual 'if the quantity of supplied goods were to increase, then the price of the goods would decrease'. According to CVS, this counterfactual is true in a world iff the price decreases in the closest worlds in which the supply increases. Following the CVS selection function, the closest worlds are those worlds in which (a) it is the case that the supply of goods increases, (b) other causes of the price (say, the demand for the good) are held constant, and (c) the laws of supply and demand hold. If we adopt current microeconomics, we should evaluate the counterfactual as true.

Although a majority of philosophers would prefer possible worlds semantics, I see another option. Following philosophers such as Pearl and Maudlin, I can also adopt a Goodmanian version of CVS.

### The Goodmanian Version of CVS

*If $\phi$ were the case, then $\psi$ would be the case* is true relative to a causal model M = $\langle$U, V, L$\rangle$ iff $\psi$ follows from the following premises:

- the event statement $\phi$,
- statements expressing that the direct causes $W_1, ..., W_n$ of the consequent (that are off-path) are in the redundancy range, and
- a set of laws L.

According to the Goodmanian version of CVS, the counterfactual 'if the quantity of supplied goods would increase, then the price of the goods would decrease' is true relative to a causal model iff the consequent ('the price of the goods decreases') follows from the following premises: first, the supply of the good increases; second, other causes of the price (for instance, the demand for the good) are held fixed; and, third, the laws of supply and demand.

Note that the two versions of CVS have several attractive features and it preserves important virtues of the interventionist approach:

1. In both versions, CVS does *not* rely on the troublesome notion of a logically possible intervention. Instead, in the case of the possible worlds version, one considers a possible world in which the antecedent is true and other direct causes take values in the redundancy range (given the actual functional relations hold between the variables, described by the law statements). In the Goodmanian version, one explicitly assumes that other causes are held constant as a premise.

2. The two versions of CVS can deal with counterfactuals whose antecedent expresses an event(-type) on which an intervention is physically impossible – such as in the conditional 'if the Big Bang had occurred differently, then the universe would have evolved differently'. As I argued in Chapter 4, these kinds of counterfactuals create severe problems for interventionists, because it is not physically possible to intervene on the Big Bang, and requiring interventions to be merely logically possible leads to evaluating these counterfactuals, counterintuitively, as false. However, CVS does not run into the same difficulties: in order to evaluate the conditional, one has to *either* go to the closest worlds in which it is true that the Big Bang occurred in the counterfactual way described in the antecedent (in the case of the possible worlds version of CVS), *or* one has to derive the consequent of the conditional from a set of premises including the statement that the Big Bang has occurred in another way than it actually did (in the case of the Goodmanian version of CVS). Let me put it more cautiously, *if* one can rely on appropriate laws describing how the world evolves given counterfactual initial conditions, then CVS can provide truth conditions for the counterfactual conditional in question.

3. CVS conforms to the claims that Woodward and other interventionists endorse with respect to causal statements and counterfactuals: (a) causal statements and counterfactual claims have meaning, (b) the meaning of causal and counterfactual statements is captured by truth-conditional semantics, and (c) in order to state these truth conditions one needs to invoke causal notions. CVS affirms these claims. Moreover, CVS is conceptually non-reductive in a similar sense as interventionist semantics is. In Chapter 7, I argued that conceptually non-reductive explications can be accepted given that they are not viciously circular, and given they satisfy the naturalist criterion and the distinction criterion of adequacy in an outstanding way (I called this the 'trade-off argument'). I think that CVS and CVS-based definitions of causal notions are not viciously circular, neither do they fail to meet the requirement of being adequate, as I will show.

4. In both versions of CVS, the role of law statements is clear (contrary to the original ambiguous remarks by Woodward and Hitchcock about the relation between the truth of a counterfactual and law statements). According to CVS, laws are required to select the closest possible worlds (in the case of the possible worlds version of CVS), and they are needed to be able to infer the consequent of the counterfactual (in the case of the Goodmanian version of CVS). If the counterfactuals concern the special sciences, then I take laws to be explicated as

non-universal laws: a statement L is a special science law iff (a) L is a system law, (b) L is quasi-Newtonian, and (c) L is minimally invariant. Ultimately, it depends on the success of this theory of laws whether statements such as 'ceteris paribus, X causes Y' and 'ceteris paribus, Y counterfactually depends on X' are informative and non-trivial.

After having introduced CVS, I will now use this semantics to explicate causal notions.

## The comparative variability theory: definitions

Now, the decisive question is whether we can explicate causal notions in a way that, on the one hand, avoids the problems of interventionism and, on the other hand, preserves the attractive features and advantages of interventionism? In order to take up this challenge, recall Woodward's definition of a direct type-level cause:

> A necessary and sufficient condition for $X$ to be a direct cause of $Y$ with respect to some variable set **V** is that there be *a possible intervention on X* that will change $Y$ (or the probability distribution of $Y$) when all other variables are held fixed at some value by interventions (Woodward 2003: 55, also cf. p. 59 as part of definition **M**).

If we abstract from the notion of an intervention (and, at this stage, we seem to have every reason to believe that this is a good idea) the interventionist definition of a direct cause boils down to counterfactual dependence between $Y$ and $X$ – in counterfactual situations in which all other causes are held fixed. This idea *need not* and *should not* be spelled out in terms of interventions. We can alternatively account for the very same idea in terms of CVS. To begin with, let me refine the most important causal notion, the notion of a direct type-level cause.

### Definition: Direct type-level cause
$X$ is a direct type-level cause of $Y$ relative to a causal model M = ⟨U, V, L⟩ (with $X$, $Y$ ∈ **V**) iff:

1. *Dependence condition*: There is some value $x_i$ of $X$ on which $Y$ counterfactually depends; that is, the following two counterfactuals are true relative to M (for i≠j, k≠l): (a) if $X = x_i$ were the case, then $Y = y_k$ would be the case; and (b) if $X = x_j$ were the case, then $Y = y_l$ would be the case.

2. *Distinctness condition*: $X = x_i$ and $X = x_j$ as well as $Y = y_k$ and $Y = y_l$ represent distinct (possible) token events.
3. *Redundancy range condition*: The appropriate worlds to evaluate the counterfactuals are worlds in which all other variables in model M take values in the redundancy range with respect to $X$ and $Y$.

Let me illustrate this definition of a direct type-level cause with a simple example. Imagine that drinking coffee is a direct type-level cause of nervousness (relative to causal model M). Suppose that model $M = \langle V, L \rangle$ consists of the following set **V** of binary variables $\{C, N, S, E\}$:

- drinking coffee is modelled by a binary variable $C$ ranging over the set of possible values {coffee; orange juice};
- being nervous is modelled by a variable $N$ ranging over {nervousness; calm};
- $S$ represents the supply of coffee with the range of possible values {coffee is supplied; orange juice is supplied} (I will use S later in order to illustrate my definition of indirect type-level causation);
- $E$ represents the fact that an important maths exam is coming up in the next hour (with $E$ ranging over the possible values {exam in an hour; philosophy lecture in an hour}).

Furthermore, the causal model M includes a set of laws **L** which determine the following relations over the set **V**: (a) an increase in drinking coffee leads to an increasing nervousness (this is a law for $C$ and $N$); (b) if coffee is supplied then coffee can be consumed (this is a law relating $S$ and $C$); and (c) maths exam coming up within the next hour increases nervousness (this law relates $E$ and $N$). This model M is represented in Figure 8.1:

According to the CVS-based notion of a direct type-level cause, $C$ is a direct cause of $N$ iff the conjunction of the following conditions is satisfied:

1. *Dependence condition*: $N$ counterfactually depends on $C$, and the corresponding counterfactuals can be evaluated by CVS (which

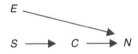

*Figure 8.1*   Coffee scenario

implies that there is at least a non-universal law specifying invariant relations between *C* and *N*): (a) if it were the case that *C* = *coffee*, then it would be the case that *N* = *nervous*, (b) if it were the case that *C* = *orange juice*, the it would be the case that *N* = *calm*.

2. *Distinctness condition*: The event-type statements *C* = *coffee* and *C* = *orange juice* as well as *N* = *nervous* and *N* = *calm* represent distinct (possible) token events (unlike, for instance, the event-types of writing 'cause' and of writing 'cau').

3. *Redundancy range condition*: The appropriate worlds to evaluate the counterfactuals are worlds in which all other variables in model M – that is, *S* and *E* – take values in the redundancy range with respect to *C* and *N* (e.g. *S* = *coffee is supplied*, *E* = *philosophy lecture in an hour*).

Let me add another example that is closer to the special sciences. Suppose that the causal claim 'the supplied quantity of a good is a direct type-level cause of the price of the good' is true relative to a causal model that includes (a) variables representing the supplied quantity of a good, the demanded quantity of a good, and the price of the good; and (b) the laws of supply and demand. According to the comparative variability theory, 'the supplied quantity of a good is a direct type-level cause of the price of the good' is true relative to this causal model iff these conditions are jointly satisfied:

1. *Dependence condition*: The price of a good counterfactually depends on the quantity supplied, and the corresponding counterfactuals can be evaluated by CVS (which implies that there is at least a non-universal law specifying invariant relations between the supply and price when other causes of the price, such as the demand for the good, are held constant).

2. *Distinctness condition*: The event-type statements representing different quantities of supplied goods as well as event-type-statements representing different prices of the good refer to distinct (possible) token events.

3. *Redundancy range condition*: The appropriate worlds to evaluate the counterfactuals are worlds in which all other variables in model M – that is, the demand for the good – take values in the redundancy range with respect to the price.

Let me emphasize that this definition of direct type-level causation is CVS-based in the sense that it permits the use of either version of CVS (i.e. the possible worlds version and the Goodmanian version) in order to evaluate the counterfactuals introduced in the dependence condition. This remark will also apply to the following explications of causal notions.

With the definition of a direct type-level cause in hand, we can proceed to define indirect causation:

### Definition: Indirect type-level cause

$X$ is an indirect type-level cause of $Y$ relative to a causal model M = $\langle$U, V, L$\rangle$ (with X, Y $\in$ V) iff:

*Path condition*: There is a directed path from $X$ to $Y$ via a set of variables $\{Z_1, ..., Z_n\}$ such that each link in the path is a direct type-level cause (i.e. $X$ is a direct type-level cause of $Z_1$, $Z_1$ is a direct type-level cause of $Z_2$, ..., $Z_n$ is a direct cause of $Y$).

*Redundancy range condition*: The appropriate worlds to evaluate the counterfactuals in question are worlds in which all other variables in the model that do not lie on the directed path from $X$ via $Z_1$, ..., $Z_n$ to $Y$ take values that are in the redundancy range.

Let me illustrate this definition of an indirect type-level cause. Recall the coffee example that I used to illustrate the definition of a direct type-level cause, which was represented by the following causal graph:

Suppose that the causal statement 'the coffee supply $S$ is an indirect cause of nervousness $N$' is true. Each link in the directed path from $S$ to $N$ via $C$ is a direct cause such that $N$ directly causes $C$ and $C$ directly causes $N$. This is in accord with the path condition of indirect type-level causation. However, $E$, representing the fact that an important maths exam is coming up in the next hour, is not on *this* directed path from $S$ to $N$ via $C$. According to the underlying causal model, $E$ is a direct cause of $N$ that lies on *another* directed path to $N$. The redundancy range condition specifies, again, the allowed assignments of values to off-path variables (in this case it is only $E$, and $E$'s value *philosophy lecture in an hour* is in the redundancy range).

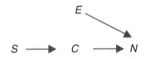

*Figure 8.2*  Indirect cause

Modally speaking, one has to specify that the definitions of direct and indirect type-level causation introduced above are definitions of *contributing* causes. Interventionists underline that contributing causes have to be distinguished modally from total or sufficient causes (cf. Woodward 2003: 50f.). These definitions characterize *contributing* causes, because whether $X$ is a cause of $Y$ does not solely depend on $X$ (as in the case of a sufficient cause) *but also on the other variables* (whose values are required to be in the redundancy range for the path from $X$ to $Y$). For this reason, we may simply equate the definition of a contributing type-level cause with the definitions of direct and indirect causation (on the type-level).

### Definition: Contributing type-level cause
$X$ is a contributing type-level cause relative to a causal model M = $\langle U, V, L \rangle$ (with $X, Y \in V$) iff $X$ is either a direct type-level cause of $Y$ or $X$ is an indirect type-level cause of $Y$ (according to the definitions above).

I defined contributing causes by pointing out that whether $X$ is a cause of $Y$ does not depend solely on $X$ *but also on the other variables* (whose values are required to be in the redundancy range for $Y$).

As opposed to contributing causes, one may now explicate the concept of a *total* or *sufficient* cause by dropping the redundancy range requirement for other variables that is assumed in the definition of contributing causes.

### Definition: Total type-level cause
$X$ is a total type-level of $Y$ relative to a causal model M = $\langle U, V, L \rangle$ (with $X, Y \in V$) iff:

1. *Dependence condition*: There is some value $x_i$ of $X$ on which $Y$ counterfactually depends; that is, the following two counterfactuals are true relative to M (for i≠j, k≠l): (a) 'if $X = x_i$ were the case, then $Y = y_k$ would be the case'; and (b) 'if $X = x_j$ were the case, then $Y = y_l$ would be the case'.
2. *Distinctness condition*: $X = x_i$ and $X = x_j$ as well as $Y = y_k$ and $Y = y_l$ represent distinct possible token events.
3. *Totality condition*: The counterfactuals (referred to in the dependence condition) are true for all possible assignments of values to the off-path variables in the causal model. In other words, the counterfactuals are true *not* because they are aptly evaluated in worlds in which all other variables in the model that do not lie on the directed

path from $X$ to $Y$ are in the redundancy range with respect to this directed path.

The totality condition expresses the idea of dropping the redundancy range requirement that is assumed in the definition of contributing causes. Note that the distinctions of direct/indirect causes and contributing/total causes can be freely combined.

Furthermore, there is a third distinction to be drawn here: the distinction between *deterministic* and *probabilistic* causes. So far, all the definitions have aimed at explicating deterministic causation. That is, if the cause $X$ were assigned a particular value, then – given certain background conditions – the effect $Y$ would *with certainty* take a specific value. However, not all causation is deterministic and it is a condition of adequacy for any theory of causation in the special sciences to account for probabilistic causation; that is, because of the fact that some causes do not bring about their effect with certainty, causes merely alter the probability of the occurrence of the effect. Although Woodward's definition **M** allows that an intervention on $X$ merely changes 'the probability distribution of $Y$' (Woodward 2003: 59), it seems to be a shortcoming of interventionist theories that probabilistic causation is not explicitly defined. This is a desideratum of the interventionist theory of causation and I will attempt to define positive and negative probabilistic causation. Let me first turn to positive probabilistic causation.

**Definition: Direct positive probabilistic type-level cause**
$X$ is a direct positive probabilistic type-level cause of $Y$ relative to M = $\langle U, V, L_{indet} \rangle$ (with $X, Y \in V$) iff:

1. *Dependence condition*: There is some value $x_i$ of $X$ on which $Y$ counterfactually depends. That is, the following two counterfactuals are true relative to M (for i≠j, k≠l): (a) 'if $X = x_i$ were the case; then the *probability increases* that $Y = y_k$ would be the case'; and (b) 'if $X = x_j$ were the case, then the *probability increases* that $Y = y_l$ would be the case'.
2. *Distinctness condition*: $X = x_i$ and $X = x_j$ as well as $Y = y_k$ and $Y = y_l$ represent distinct possible token events.
3. *Redundancy range condition*: The appropriate worlds to evaluate these counterfactuals are worlds in which all other variables in the model take values that are in the redundancy range with respect to $X$ and $Y$.

Let me add a remark: according to CVS, the evaluation of counterfactuals involving probabilities (such as the conditionals (a) and (b) in the dependence condition) requires indeterministic causal models; that is, models in which the relations between the variables are specified by probabilistic generalizations $L_{indet}$ (for elaborated work concerning this issue see, for instance, Pearl 2000: ch 7: 201–58; Maudlin 2007: 21–34).

We are now in a position to define negative or counter-acting probabilistic causation.

### Definition: Direct negative probabilistic type-level cause
$X$ is a direct negative probabilistic type-level cause of $Y$ relative to M = $\langle$U, V, $L_{indet}\rangle$ (with $X, Y \in$ V) iff:

1. *Dependence condition*: There is some value $x_i$ of $X$ which $Y$ counterfactually depends upon. That is, the following two counterfactuals are true relative to M (for i$\neq$j, k$\neq$l): (a) 'if $X = x_i$ were the case, then the *probability decreases* that $Y = y_k$ would be the case'; and (b) 'if X = $x_j$ were the case, then the *probability decreases* that $Y = y_l$ would be the case'.
2. *Distinctness condition*: $X = x_i$ and $X = x_j$ as well as $Y = y_k$ and $Y = y_l$ represent distinct token events.
3. *Redundancy range condition*: The appropriate worlds to evaluate the counterfactuals are worlds in which all other variables in the model take value in the redundancy range with respect to $X$ and $Y$.

Based on the definitions of *direct* positive and negative probabilistic causation, one can continue to define *indirect* probabilistic causation analogously to indirect deterministic causation.

### Definition: Indirect probabilistic type-level cause
$X$ is an indirect probabilistic type-level cause of $Y$ relative to a causal model M = $\langle$U, V, $L_{indet}\rangle$ (with $X, Y \in$ V) iff:

1. *Path condition*: There is a directed path from $X$ to $Y$ via a set of variables $\{Z_1, ..., Z_n\}$ such that each link in the path is a direct *positive or negative probabilistic* type-level cause (i.e. $X$ is a direct *positive or negative probabilistic* type-level cause of $Z_1$, $Z_1$ is a direct *positive or negative probabilistic* type-level cause of $Z_2$, ..., $Z_n$ is a direct *positive or negative probabilistic* type-level cause of $Y$).
2. *Redundancy range condition*: The appropriate worlds to evaluate the counterfactuals in question are worlds in which all other variables in the model that do not lie on the directed path from $X$ via $Z_1, ..., Z_n$ to $Y$ take values that are in the redundancy range with respect to this path.

All of the definitions introduced up to this point concern type-level causation. Type-level causation does not exhaust the varieties of causal phenomena: many philosophical disputes regard actual causation, i.e. causally related actual events, and, to account for actual causation has been considered to be a condition of adequacy for any theory of causation in the special sciences on pp. 11–21. Building on the interventionist approach to actual causation (cf. Woodward 2003; Hitchcock 2001; Halpern and Pearl 2005), one can define direct and indirect actual causation in the framework of Comparative Variability theory as follows:

### Definition: Direct actual cause
$X$ is a direct actual cause of $Y$ relative to a causal model M = $\langle$U, V, L$\rangle$ (with $X, Y \in$ V) iff:

1. *Actuality condition*: The event statements $X = x_i$ and $Y = y_k$ are true in the *actual* world.
2. *Distinctness condition*: $X = x_i$ and $Y = y_k$ represent distinct *actual* token events, $X = x_i$ and $Y = y_l$ represent distinct *possible* token events.
3. *Dependence condition*: $Y$ counterfactually depends on $X$ such that the following two counterfactuals are true relative to M (for i≠j, k≠l): (a) if $X = x_i$ were the case, then $Y = y_k$ would be the case; and (b) if $X = x_j$ were the case, then $Y = y_l$ would be the case.
4. *Redundancy range condition*: The appropriate worlds to evaluate the counterfactuals are worlds in which all other variables in the model are in the redundancy range with respect to $X$ and $Y$.

Let me draw once more on the coffee illustration. Suppose that 'the event that Anna drank coffee this morning at 10am is the actual cause of her nervousness at 11am' is true. According to my definition of actual causation, the truth conditions for this causal claim are:

1. *Actuality condition*: Anna actually instantiates the event-types $C = coffee$ today at 10am and $N = nervous$ at 11am.
2. *Distinctness condition*: $C = coffee$ and $N = nervous$ represent distinct *actual* token events, $C = orange juice$ and $N = calm$ represent distinct *possible* token events.
3. *Dependence condition*: $N$ counterfactually depends on $C$ such that the following two counterfactuals are true (of Anna) relative to M (for i≠j, k≠l): (a) if it were the case that $C = coffee$ at 10am, then it would be the case that $N = nervous$ 11am; (b) if it were the case that $C = orange juice$ at 10am, then it would be the case that $N = calm$ at 11am.

4. *Redundancy range condition*: the appropriate worlds to evaluate the counterfactuals are worlds in which all other variables in the model are in the redundancy range with respect to $C$ and $N$.

Analogously to the definitions of type-level causation, I can define *indirect* actual causation.

### Definition: Indirect actual cause
$X$ is an indirect actual cause of $Y$ relative to a causal model M = $\langle$U, V, L$\rangle$ (with $X, Y \in$ V) iff:

1. *Path condition*: There is a directed path from $X$ to $Y$ via a set of variables $\{Z_1, ..., Z_n\}$ such that each link in the path is a direct actual cause (i.e. $X$ is a direct actual cause of $Z_1$, $Z_1$ is a direct actual cause of $Z_2$, ..., $Z_n$ is a direct actual cause of $Y$).
2. *Redundancy range condition*: The appropriate worlds to evaluate the counterfactuals in question are worlds in which all other variables in the model that do not lie on the directed path from $X$ via $Z_1$, ..., $Z_n$ to $Y$ take values that are in the redundancy range with respect to this path.

Let me review what has been accomplished in this section. I have defined various causal notions on the basis of CVS: direct and indirect type-level causation, contributing and total type-level causation, positive and negative probabilistic causation, and actual causation. I will now turn to the question whether these fruit of the comparative variability theory of causation pass the test of being an adequate theory of causation for the special sciences.

## An adequate explication of causation

Does the comparative variability theory meet the criteria of adequacy for an account of causation in the social sciences, and, hopefully, for the special sciences in general? The comparative variability theory of causation is acceptable as an adequate explication of causation in the context of these disciplines only on the condition that it satisfies the naturalist criterion and the distinction criterion. Next, I will investigate whether the comparative variability theory conforms to the sub-criteria of adequacy that fall under the naturalist criterion (i.e. characteristic kinds and features of causation in the special sciences). The result of this investigation will be positive, and more successful than for the interventionist theory. I will then turn to the distinction criterion of an adequate analysis. I will show that the comparative variability theory is able to describe correctly and naturally the causal structures that trouble many other theories of causation as counter-examples. In order to argue for this claim, I will discuss a paradigmatic scenario of pre-emption and the controversial assumption that causation is a transitive relation. I will conclude that the comparative variability theory of causation promises to be a good candidate for an adequate explication of causation in the special sciences.

## The naturalist criterion of adequacy

Chapter 1 presented a list of typical and characterizing kinds of causation and features of causation in the social sciences, and in the special sciences in general. I gave the label 'naturalist' to this list of kinds and features of causation, because scientists refer to these kinds of causation and scientists assign these features to various kinds of causation. By contrast, it does not matter for my project whether speakers in non-scientific everyday contexts refer to the same kinds and features of causation. In order to provide an adequate explication of causation in the special sciences, one has to present a definition for various kinds of causation, and one is required to explain why causation has the typical features that scientists (explicitly or implicitly) assign to it. In other words, one has to meet the naturalist criterion if one wants to provide adequate explication of causation.

Let me now go through this list step by step and argue that the comparative variability theory of causation can capture the kinds of causation and explain the features of causation that the naturalist criterion encompasses.

1. *Type-level and token-level causation*: The special sciences, by and large, are interested in type-level causes as well as in actual token-causes. Type-level causation and actual causation differ with respect to the relata standing in causal relations – type-level causation relates event-types, and actual causation relates event-tokens.

Analogously to interventionist theories of causation, the comparative variability theory is able to define various notions of type-level causation (e.g. direct, indirect, total, contributing causation, probabilistic, and indeterministic causation). I can also account for the notion of actual causation. In a similar way to the interventionist theory, it is explicated derivatively of type-level causation in the sense that event-types (represented by statements of the canonical form X = x) are taken to be instantiated by actual objects. The actual instantiation of such an event-type is an actual event, and actual events are the relata of actual causation. In sum, the comparative variability theory captures these kinds of causation.

2. *Deterministic and probabilistic causation*: According to some of examples of causal statements in the special sciences, some cases of causation are deterministic, while other causes are probabilistic (i.e. the causes make a probabilistic difference to the effect).

Deterministic and probabilistic causation is explicitly defined in the comparative variability framework (negative, positive, direct, and indirect probabilistic causation). As observed, Woodward does indeed account for probabilistic causation in his definitions of (in)direct contributing causation by postulating that 'a possible intervention on X that will change Y (or *the probability distribution of Y*)' (Woodward 2003: 59, emphasis added). Nonetheless, probabilistic causation appears to be somewhat of a side issue for Woodward. Although there is no principle advantage for the comparative variability theory, one might still insist that comparative variability theory does, in fact, take better care of probabilistic causation than do interventionists: in the framework of the comparative variability theory, one makes explicit the distinction between deterministic and probabilistic causation by defining various notions of probabilistic causation (i.e. direct and indirect, positive or negative probabilistic causation). Furthermore, the comparative variability theory assigns a greater value to the probabilistic kinds of causation by ranking the definition of probabilistic causes as a criterion of adequacy for a theory of causation for the special sciences.

3. *Multiple causes and degrees of causal influence*: Social scientists hold that real-life cases of causation usually involve multiple causes of a phenomenon, and these multiple causes contribute in differing degrees to the effect.

The most natural interpretation of 'multiple causes' is to say that a phenomenon has a number of contributing causes. Since the comparative variability theory provides a definition of the notion of contributing causation, the theory can easily account for a scenario in which an effect has multiple (type-level or actual, direct and/or indirect, deterministic and/or probabilistic) contributing causes. Additionally, the comparative variability theory can neatly account for the fact that causal statements in the special sciences often assign differing degrees of influence to a cause. Deterministic and probabilistic laws represent degrees of causal influence. A paradigmatic example can be taken from the causal modelling literature: degrees of causal strength are usually represented as coefficients in structural equations (Cartwright 1989; Pearl 2000; Spirtes et al. 2000; Hoover 2001). Thus, the comparative variability theory accounts for this kind of causation also.

4. *Modal relation*: Causes influence, enforce, bring about or produce their effects. In other words, some kind of modal connection obtains among cause and effect.

One of Hitchcock's (2007: 57f.) difficult problems with causation consists in explaining the 'modal force' of causal relations in a way that distinguishes causal relations from accidental correlations. The intuition of attributing modal force to causation can be illustrated by an example: I believe that turning the switch is a cause of the bomb blowing up. It is not a mere correlation between turning the switch and the detonation of the bomb. The intuition is that turning the switch enforced, brought about, and produced the detonation. It is the assumption that a modal connection obtains between turning the switch and the detonation that kept me from turning the switch because I wanted to avoid the detonation. The question is: How shall we make sense of this intuition that causes are somehow modally connected to their effects? If the modal force is supposed to be interpreted as the claim that a single cause is always *sufficient* for its effect(s) in all situations, then this can be met with a straightforward denial: most explications of causal notions concern contributing causes that are not sufficient for their effect, because their causal influence depends – by definition – on other causes of the effect.

In the debate on causation, the question whether causes bring about and produce their effects is regarded as a *metaphysical* dispute. There are many metaphysical options that can be assessed. Humeans issue a straightforward denial that causation is a modal (i.e. necessary) connection between events: causation is (or supervenes on) nothing but a regular pattern of contingent events. Humeans maintain that – even though we do not rely on modal facts and modal connections among events – one is *still* able to distinguish accidental correlations from regularities by virtue of which causal statements are true. However, there are various alternative ways to understand how causes 'produce' their effects: a non-modal reading according to which causation is a specific kind of physical process, and a non-Humean modal reading according to which causes necessitate their effects.

According to the non-modal reading, causation is interpreted as the possession or exchange of conserved quantities (cf. Dowe 2000). Another way to approach the productivity of causation is to assume that production is conceptually and metaphysically primitive; it is not further explicable sense (cf. Machamer 2004; Bogen 2005; Moore 2009). This kind of primitivism can be subsumed under the modal reading of production. Yet, one need not be a primitivist in order to subscribe to the modal reading. For instance, Armstrong (1983) famously claims that causal relations and laws of nature are modally laden entities. According to Armstrong, a law is the second-order universal of necessitation relating first-order universals and a causal relation is the instantiation of a law of nature. Armstrong's relation of necessitation is itself a modal relation.

Now the question is whether I have to make a decision in favour of one of these metaphysical Humean and anti-Humean alternatives. I do not think this is the case and I hope that I am capable of remaining neutral. The comparative variability approach seems to be in agreement with interventionist theory that there is a sort of *minimal* modal connection between cause and effect that is entailed by the theories of causation. Woodward's approach and my approach provide a strategy to distinguish nomic and causal relations from merely accidental correlations by referring to the invariance of law statements. As I argued in Chapter 5, this distinction is a distinction in kind. However, this strategy to distinguish nomic and causal relation from merely accidental correlations is metaphysically neutral with respect to the Humean and anti-Humean claims presented above. Consequently, it cannot solve the metaphysical riddles. However, use of the invariance strategy to distinguish nomic and causal relations from merely accidental correlations provides something one might call a minimal account of the modal connection. Moreover, my comparative variability theory and Woodward's approach also suggest a minimal understanding of the modal force of causation in terms of the variably strict necessity that is a feature of (interventionist) counterfactual conditionals (cf. Lewis 1973a: 13–19). In a sense, this necessity is a nomological necessity because it is grounded in the law statements (and structural equations) that describe the causal structure at hand. However, this minimal reading is very limited because the question of modal force reappears with respect to the laws. The truth-makers of law statements might, among other metaphysical views, be Humean regularities, physical processes involving conserved quantities, a kind of production that is conceptually and metaphysically primitive, or Armstrong's relation of necessitation. I do not want to take sides in this metaphysical dispute about laws. It seems to be an entirely open question as to which metaphysical option best fits my account of non-universal law statements. This is not a vice of my account. In contrast, I think it a virtue of my theory of causation that it is *not* committed to a particular view of the metaphysics of causation.[7] However, I'm still able to claim that there is a minimal modal connection linking cause and effect because (a) I presented a strategy to distinguish nomic relations from merely accidental correlations (see Chapter 5), and (b) my view of causation entails that causes are nomologically sufficient for their effects (relative to a given causal model that partly consists of a set of deterministic law statements, and relative to a set of variables that are held constant).

5. *Time-asymmetry of causation*: Causes precede their effects in time, and not vice versa.

In Chapter 6, I argued that interventionists fail to account for the time-asymmetry of causation. The reason for this failure is that the interventionist argument for the claim that causation is time-asymmetric, the open-systems argument, is not sound. I presented an alternative to the open-systems argument: the statistical mechanical approach to explain the time-asymmetry of causation (cf. Albert 2000; Kutach 2007; Loewer 2007, 2009). According to this approach, time-asymmetric relations can (at least, in principle) be explained by the same assumptions that are used to explain the time-asymmetric second law of thermodynamics: this explanation draws, among other things, on the time-symmetric laws of statistical mechanics, and on the assumption that the early state of the universe was a state of low entropy (the past hypothesis). Another way to put it is: causal facts and entropic facts supervene on the same fundamental nomic and particular facts. From our perspective, this approach is attractive: it does not rely on (merely logically possible) interventions. A proponent of the comparative variability theory of causation can use the statistical-mechanical (SM) approach by Albert, Kutach, and Loewer in order to explain the existence of time-asymmetric, non-universal laws (Loewer 2009). These time-asymmetric laws are commonly used in causal modelling, and by interventionists in particular: these laws usually take the form of structural equations. Once the existence of (non-fundamental) time-asymmetric laws is explained and granted, it is obvious that counterfactuals that are evaluated with respect to these time-asymmetric laws are also time-asymmetric, because *either* these time-asymmetric laws obtain in the closest possible worlds (in the case of the possible worlds version of CVS), *or* these time-asymmetric laws are used to derive the consequent of the counterfactual in question (as in the case of the Goodmanian version of CVS). I promote a counterfactual theory of causation according to which causation is also time-asymmetric. Consequently, I can use the SM-account as an external explanation of why causation has certain features. Therefore, I am entitled to claim that *if* the comparative variability theory is supplemented with the SM-account, I can then explain why causation is time-asymmetric.

6. *Causal asymmetry*: The causal relation is asymmetric; that is, if X causes Y, then Y does not cause X.

In Chapter 2, I diagnosed that interventionists fail to account for the features of time-asymmetry and causal asymmetry in a straightforward way. Their promised explanation seemed to be the open-systems argument. This argument was supposed to explain why causation possesses certain typical features. However, the open-systems argument is not sound, as I argued in Chapter 6. This is a problematic result for interventionists. Yet, I can use a result of the earlier discussion of the interventionist theory in order to account for the feature of causal asymmetry (i.e. effects depend on their causes but not vice versa). I have argued that causal asymmetry does not imply the time-asymmetry of causation, but the time-asymmetry of causation implies causal asymmetry. If this is true, then I have an indirect argument for causal asymmetry – if the time-asymmetry of causation implies causal asymmetry. If one accepts the statistical-mechanical explanation for the time-asymmetry of causation, then causation has the feature of causal asymmetry. Let me state this claim in a more detailed way.

I assumed that if $X$ causes $Y$ then $Y$ asymmetrically depends on $X$, because $Y$ counterfactually depends on $X$. This is the core of any counterfactual theory of causation and, therefore, it is a basic assumption of comparative variability theory. A challenge for any counterfactual theory of causation is to rule out backtracking counterfactuals; that is, counterfactuals according to which the cause $X$ depends on the effect $Y$. As I argued in Chapter 6, the SM-account provides a reason to 'filter out' backtracking counterfactuals, if we believe that causes precede their effects in time. So, I propose to account for causal asymmetry in two steps: first, we claim that the effect $Y$ counterfactually depends on cause $X$; second, we rule out backtracking counterfactual dependence by relying on the SM-account for the time-asymmetry of causation. Hence, if we combine comparative variability theory and the SM-account, then we can provide an explanation for the feature of causal asymmetry.

7. *Distinction of spurious correlation and genuine causation*: There is an important distinction between (a) the case of a correlation between $X$ and $Y$ due to a common cause $Z$, and (b) the case of a causal relation between $X$ and $Y$ (i.e. either $X$ causes $Y$ or vice versa).

In Chapter 2, I used the classic barometer example as an illustration. Now I will simply discuss the abstract case for brevity's sake. Suppose that $X$ and $Y$ are positively correlated, and it is neither the case that $X$ causes $Y$ nor that $Y$ causes $X$. In fact, $X$ and $Y$ are the effects of a common cause $Z$. The intuition of the example is that $X$ and $Y$ are *merely correlated*, while there is a *genuinely causal relation* between the common cause and its effects $X$ and $Y$. A common cause scenario is represented In Figure 8.3.

I think the comparative variability theory can describe the common cause scenario correctly, because it is neither the case that $X$ counterfactually depends on $Y$ if the common cause $Z$ is held fixed, nor does $Y$ counterfactually depend on $X$ if the common cause $Z$ is held fixed. Thus, according to my approach, it is neither the case that X causes Y, nor that Y causes X. However, $X$ and $Y$ counterfactually depend on the common cause $Z$. Thus, the comparative variability theory is able to distinguish spurious correlations from genuine causation in a common cause scenario.

8. *Context*: Many of the paradigmatic examples of causal statements in the special sciences are (implicitly) believed to hold only on the assumption that other variables are held fixed. Economists and philosophers often refer to this feature with the expression 'ceteris paribus'.

*Figure 8.3*  Structure of a common cause scenario

I reconstructed and adopted Marshall's key idea about causation in economics by claiming that typical causal statements are of the form 'ceteris paribus, X causes Y'. I added the claim that causation is supposed to be explicated in terms of counterfactual dependence. Combining the two elements – that is, ceteris paribus clauses and the counterfactual theory of causation – gives rise to the comparative variability theory of causation according to which, roughly, X causes Y iff Y counterfactually depends on X when ceteris paribus conditions (in the sense of Schurz's comparative reading of the ceteris paribus clause) obtain or, equivalently, when other factors are held fixed in an appropriate way. The fact that context matters in the sense that causal statements are believed to be true only on the assumption that other variables are held fixed is explicitly built into the CVS and the definitions of causal notions provided by the comparative variability theory. Ceteris paribus conditions are also mirrored in the definitions of contributing type-level and actual causes which are defined with respect to a redundancy range of other (off-path) variables. Moreover, the holding-fixed of the causal context is also essentially employed for characterizing the invariance of the non-universal laws that describe the relations among the variables in a causal model (see Chapter 5). These laws are crucial building blocks of the possible worlds version and the Goodmanian version of CVS. In sum, the comparative variability approach performs extremely well in coping with the context-criterion of adequacy.

9. *No-universal-laws requirement*: Since there are no universal laws in the social sciences (and, possibly, in the special sciences in general), an explication of causation for these disciplines cannot refer to such laws.

The comparative variability theory of causation does not rely on universal laws in order to state the truth conditions of causal statements. Instead, it refers to a weaker kind of generalization in the semantics for counterfactuals. I take laws to be explicated as non-universal laws (see Chapter 5): a statement L is a special science law iff (a) L is a system law, (b) L is quasi-Newtonian, and (c) L is minimally invariant*. These generalizations do not satisfy the characteristics of laws as philosophers traditionally conceive them. However, despite lacking several received features of lawhood, non-universal laws can be distinguished from merely accidentally true, universal statements, and they qualify as true empirical statements. I conclude that the comparative variability theory of causation satisfies the no-universal-laws requirement to a considerable degree.

Let me briefly take stock. I have confronted the comparative variability theory of causation with nine sub-criteria of adequacy that are subsumed under the naturalist criterion for the explication of causation in the special sciences. It seems to me that the comparative variability approach conforms to all of the nine sub-criteria. If this is correct, my theory of causation is more successful than the Woodward's interventionist theory. Comparative variability theory offers an account for those sub-criteria of the naturalist criterion that caused difficulties for interventionists: comparative variability theory accounts for the features of time-asymmetry and causal asymmetry, and the no-universal-laws requirement. Taking these results into account, the comparative variability theory of causation conforms to the naturalist standard in a promising way.

### The distinction criterion of adequacy

In Chapter 3, I discussed two kinds of counter-examples that play a central role in the literature on causation: I presented various scenarios of pre-emption (*early* pre-emption, *late* pre-emption, *trumping* pre-emption, symmetric over-determination), and scenarios that reveal the intransitivity of causation. The result of this discussion is that Woodward's interventionist theory is able to refute these scenarios as counter-examples. Now, it is time to test the comparative variability theory of causation. I believe that the test should be easily passed, because comparative variability theory shares all the features with interventionist theories that guarantee the success of the latter. For this purpose, I will face these alleged counter-examples in two steps: first, I will discuss only early pre-emption for the case of comparative variability theory (and I will point out to the reader that the other pre-emption scenarios can be described by means of a similar strategy). Second, I will discuss the boulder scenario as a counter-example to the transitivity of causation.

So, let me first turn to early pre-emption. Imagine, once more, the following situation (cf. Lewis 1986a, 2004; Hall 2004): two assassins A and B have conspired to shoot the dictator of their country. They broke into a building across the street from where the dictator gives his speech. On the second floor, the assassins take aim, side by side. Both assassins are excellent and reliable marksmen (i.e. we model the scenario as a deterministic one such that if one assassin shoots he hits his target). Assassin A fires a shot and the dictator is killed in the middle of his speech. B desisted from firing his gun, because he saw that A was pulling the trigger. B could have backed up A, but he did not. As in Chapter 2, I represent the situation with the following variables and their values (cf. Woodward 2003: 77–9; Halpern and Pearl 2005: 861f.):

- *Exogenous variables $U_i$*: $U_A$ represents the motivation of assassin A to kill, to plan it and to take actions that are necessary preparations for the assassination etc.; respectively $U_B$ represents the same motivation for B. The fact that A is motivated to kill is represented by $U_A = 1$; not being motivated is represented by $U_A = 0$. Respectively for $U_B$. $U_A$ and $U_B$ are supposed to take the value 1.
- *Endogenous variable A*: *A* represents the shot of assassin A. Shooting is represented by $A = 1$; not shooting by $A = 0$.
- *Endogenous variable B*: *B* represents the shot of assassin B. Shooting is represented by $B = 1$; not shooting by $B = 0$.
- *Endogenous variable D*: *D* represents the state of the dictator. $D = 1$ represents being fatally hit by a bullet; $D = 0$ represents not being hit by a bullet (i.e. survival).

We represent the causal structure of the scenario in Figure 8.4.

The arrow from *A* to *B* models that whether A shoots has a causal influence on whether B shoots. The causal information contained in Figure 8.4 may be expressed as follows: given the $U_i$ take the value 1 and, if it were the case that either $A = 1$ or $B = 1$, then it would be the case that $D = 1$. And, if it were the case that $A = 0$ and $U_B = 1$, then it would be the case that $B = 1$. Otherwise – that is, for all other combinations of the values of the variables – it is the case that $D = 0$.

What is the challenge of early pre-emption to the comparative variability theory? The underlying assumption in the scenario is that A's shot is the actual cause of the dictator's death. So, according to any counterfactual theory of causation, the counterfactual conditional 'if A had not shot, then the dictator would not have died' has to be true. Yet, this counterfactual seems to be false in the scenario of early pre-emption, because assassin B would have shot and killed the dictator if A had

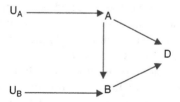

*Figure 8.4*  Early pre-emption scenario

desisted from firing his gun. In other words, the counterfactual is false because A had a back-up, B. However, the scenario of early pre-emption is only at first sight a counter-example to the comparative variability theory, because all conditions of our definition of actual causation are satisfied and, therefore, A's shot ($A = 1$) is an actual cause of the dictator's death ($D = 1$). Let me explain by going through the conjunction of the four conditions of my comparative variability definition of actual causation individually. Recall the definition of direct actual causation according to the comparative variability theory:

**Direct actual cause**
$X$ is a direct actual cause of $Y$ relative to a causal model M = $\langle$U, V, L$\rangle$ (with $X$, $Y$ in **V**) iff:

1. *Actuality condition*: The event statements $X = x_i$ and $Y = y_k$ are true in the *actual* world.
2. *Distinctness condition*: $X = x_i$ and $Y = y_k$ represent distinct *actual* token events, $X = x_j$ and $Y = y_l$ represent distinct *possible* token events.
3. *Dependence condition*: $Y$ counterfactually depends on $X$ such that the following two counterfactuals are true relative to M (for i≠j, k≠l): (a) if $X = x_i$ were the case, then $Y = y_k$ would be the case; and (b) if $X = x_j$ were the case, then $Y = y_l$ would be the case.
4. *Redundancy range condition*: The appropriate worlds to evaluate the counterfactuals are worlds in which all off-path variables (i.e all other variables because we are assuming that $X = x_i$ is a direct actual cause of $Y = y_k$) in the model are in the redundancy range with respect to the path from $X$ to $Y$.

The *actuality condition* is satisfied because A's shot and the dictator's death are actual events. The *distinctness condition* is also satisfied because A's shot and the dictator's death are spatio-temporally distinct actual events. The *dependence condition* is also satisfied: whether $A = 1$ actually causes the death of the dictator ($D = 1$) depends on *holding fixed* other variables at their (actual) values that are not located on a causal chain (or a directed path) between the shot and the death. The behaviour of assassin B is not on such a directed path from A's shot to the death of the dictator. In order to refute the counter-example, one has to rephrase the counterfactual in question as a '*conditional* counterfactual dependence' (Field 2003: 452) with additional information in the antecedent as CVS requires: 'if it were the case that A did not shoot *and B did not fire his gun* (as in the actual situation), then the dictator would not have died' is true according to the scenario of early pre-emption. In other words, the *dependence condition* is satisfied. The *redundancy range condition* is also satisfied, because (given that $A = 1$) holding the value of $B$ fixed at its actual value does not change the actual value of $D$; namely, $D = 1$.

Therefore, the definition of actual causation provided by the comparative variability theory is able to describe situations of early pre-emption correctly and, consequently, this kind of situation is not a counterexample. In Chapter 2, I presented and discussed three other important scenarios of pre-emption (the scenarios of late pre-emption, trumping pre-emption, and symmetric overdetermination). However, I do not want to go through a lengthy discussion of the other pre-emption cases. I hope the reader agrees with me that the remaining cases can be described successfully by claiming that the actual effect depends on the actual cause under the assumption that the pre-empted cause (or the symmetrically over-determining cause) is held fixed on a value within the redundancy range. Concerning this strategy, I entirely agree with Woodward's treatment of these other scenarios of pre-emption.

Let me finally turn to the scenario suggesting that causation is not transitive (cf. Hall 2004) that I presented in Chapter 2. Imagine, again, the following situation: a boulder is dislodged somewhere in the mountains (event $c$) while Mrs A is hiking close by. Hiker A sees the boulder, she ducks sufficiently early and is able to hide in a cave (event $d$). Hiding in this cave guarantees her survival (event $e$). Unless she had hidden, she would have definitely been killed. Recall that Lewis's theory of causation explicitly claims that causation is transitive. If one agrees with Lewis that causation is transitive, then one has to conclude that the boulder being dislodged causes A's survival. Yet, this does not seem to be a proper description of the causal history in the example. The decisive question is whether the comparative variability theory has to admit that causation is transitive. As in Chapter 2, I represent the situation by choosing the following variables and their values:

- *Exogenous variables $U_i$*: A variable $U_1$ might represent the mental states of hiker A (e.g. that she has the desire to hike and to survive her hiking trip, her belief that it is a perfect day to go hiking and so on). $U_1 = 1$ means that hiker A is in those mental states; $U_1 = 0$ means that she is not. We may also consider a variable $U_2$ that represents some relevant geological and climatic factors that cause the boulder to roll down the mountain. $U_2 = 1$ means that these factors are present; $U_2 = 0$ represents the fact that they do not. The $U_i$ are set to the value 1.
- *Endogenous variable $B$*: $B$ represents the boulder. $B = 1$ means that the boulder is dislodged; $B = 0$ means that it is not.

- *Endogenous variable D*: The variable $D$ represents the event of the hiker's ducking. $D = 1$ represents the fact the hiker ducks; $D = 0$ represents that she does not.
- *Endogenous variable S*: The variable $S$ stands for the survival of the hiker. $S = 1$ means that she survives; $S = 0$ means that she does not.

Figure 8.5 represents our scenario.

The causal information contained in Figure 8.5 may be expressed as follows: given that the $U_i$ take the value 1:

- if it were the case that $B = 1$ and $D = 1$, then it would be the case that $S = 1$;
- if it were the case that $B = 0$ or $D = 1$, then it would be the case that $S = 1$;
- if it were the case that $B = 0$ and $D = 0$, then it would be the case that $S = 1$;
- if it were the case that $B = 1$ and $D = 0$, then it would be the case that $S = 0$.

So, am I forced to admit that the boulder going down the hill is the actual cause of the survival of our hiker? Is the comparative variability definition of actual causation committed to this causal statement? I do not think that I have to accept transitivity. I can simply accept the boulder scenario as counter-example to the assumption that causation is a transitive relation. For a proponent of the comparative variability theory, it seems natural to join interventionists in accepting the boulder scenario as a counter-example to the claim that causation is transitive. The comparative variability theory and interventionist theory do not have to care about transitivity, because – unlike in Lewis's theory – it is not required in their definitions of (actual) causation that causation has to be transitive. Nevertheless, this amounts to an obligation for me. I have to show that, if one adopts the comparative variability theory, the description of the boulder scenario is in perfect accord with our

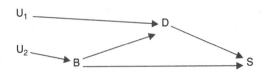

*Figure 8.5*   Boulder scenario

intuitions about the scenario: the boulder going down the mountain is *not* the actual cause of the hiker's survival, because our definition of actual causation is not satisfied for these two events. Let us examine whether this is true.

The *actuality condition* is satisfied because the boulder going down the mountain and the survival of our hiker are actual events. The *distinctness condition* is satisfied, because the boulder going downhill and the hiker's survival are spatio-temporally distinct, actual events. What about the *dependence condition*? The following counterfactual seems to be true in our scenario: 'if it were the case that $B$ took the counterfactual value 0, and we hold $D$ fixed at its actual value $D = 1$, then the value of $S$ would not change (i.e. $S = 1$)'. If that is correct, then there is no counterfactual dependence of the survival on the falling of the boulder because the following two counterfactuals are true: first, if (counterfactually) there were no boulder *and* our hiker ducked, then it would be the case that the hiker survives. And second, if there were a boulder (as there actually is) *and* our hiker ducked, then it would be the case that the hiker survives. Therefore, the *dependence condition* is *not* satisfied. It follows that $B = 1$ cannot be an actual cause of $S = 1$. One might wonder whether the *redundancy range condition* has been violated. This is not the case. The counterfactuals in question hold $D$ fixed at its actual value 1. $D = 1$ is in the redundancy range, because the fact that $D = 1$ does not alter the actual value of $S$ (i.e. $S = 1$) if $B$ takes its actual value (i.e. $B = 1$). Thus, the comparative variability theory of causation is able to state clearly that the boulder is *not* an actual cause of A's survival. According to the comparative variability theory, the definition of actual causation is not satisfied in the boulder scenario. Therefore, an advocate of comparative variability theory can give a straightforward denial that causation is transitive.

## Conclusion

I have argued in this chapter that the comparative variability theory of causation can replace the original interventionist theory. Comparative variability theory aims to exploit two essential features of interventionism: first, the claim that the truth conditions of causal statements can be stated in terms of non-backtracking counterfactual dependence between effect and cause; and, second, the claim that counterfactual dependence is somehow conditional on holding fixed *other* causes of the putative effect; that is, dependence is conditional on the qualification by a comparative ceteris paribus clause (cf. Hitchcock 2001;

Field 2003; Cartwright 2007; Menzies 2007; Heckman 2008, for related approaches). I introduced two versions of comparative variability semantics (CVS) for counterfactuals: the possible worlds version and the Goodmanian version. CVS does not face the difficulties faced by interventionists when it comes to evaluating counterfactuals with a physically impossible antecedent such as the notorious counterfactual 'if the Big Bang had occurred differently, then ...'. I provided various explications of causal notions based on the semantics introduced previously, and have shown that the comparative variability theory satisfies the criteria of adequacy for an explication of causation in the special sciences in a way that is superior to Woodward's interventionist theory. The discussion of the naturalist criterion has especially shown that comparative variability theory can account for all of the sub-criteria of the naturalist criterion – including sub-criteria that I discussed as desiderata of the interventionist theory (such as the features of causal asymmetry and time-asymmetry, and the requirement of non-universal laws). However, the explanation of the feature of time-asymmetry cannot be given by the comparative variability theory alone. The explanation is external: it is provided by the SM-account. I draw two conclusions from these results: First, the comparative variability theory of causation is a serious candidate for an adequate explication of causation in the special sciences, because it meets all of the criteria of adequacy. For this reason, we should prefer it to Woodward's interventionist theory of causation.

Second, the success of the comparative variability theory supports the *dispensability argument* developed in Chapter 4. Recall that to posit an entity (such as an intervention) is dispensable if a similar and at least equally powerful rival theory that does not posit this entity is equally or more successful. For the reason that (a) the successful comparative variability theory shares crucial features with interventionist theory (most importantly, the feature of understanding causation in terms of counterfactual difference-making, and the feature of holding other causal factors fixed), and for the reason that (b) comparative variability theory does not use the notion of an intervention, I feel justified in believing that interventions are *dispensable* for stating the truth conditions of causal statements.

In Chapter 9, I will explore the consequences of the comparative variability theory of causation, and show the impact that my theory of causation has on the debates concerning causal explanations, mechanisms, and dispositions.

# 9
# Consequences

In Chapter 8, I presented my approach to causation in the special sciences: the comparative variability theory of causation. I argued that my account should replace the interventionist theory of causation because it proves to be superior in meeting the criteria of adequacy for explicating causation in the special sciences. Further, comparative variability theory avoids the semantic and metaphysical problems connected with Woodwardian interventions, because my theory does not rely on interventions in order to explicate causation (and non-universal laws). In this final chapter, I wish to show how my approach has impact on other topics in the philosophy of science. I will focus on three topics that take centre stage in general, and especially philosophy of science: theories of explanation, representations of mechanisms, and the conditional analysis of dispositions. I will conclude this chapter by summarizing the results of the book and by providing an outlook on future research that is suggested in the light of these results.

## Causal explanation

According to Hempel's covering law model of scientific explanation, an event (the explanandum) is explained iff its occurrence is to be *expected* on the grounds of specific initial conditions and laws of nature (the explanans) (Hempel 1965). In the event that the relevant laws are universal and deterministic, the explanandum is to be expected with certainty, because the explanandum deductively follows from the laws and initial condition. This idea is expressed in the well-known deductive nomological model of explanation. If the laws are probabilistic, Hempel's inductive-statistical model of explanation requires that one should at least be able to assign a high subjective probability to the occurrence of explanandum event.

That is, given the probabilistic laws and specific initial conditions (which do not leave out relevant facts), one should expect that the explanandum event is likely to take place. Let me illustrate the covering law approach with an example that Woodward and Hitchcock discuss:

> Consider an explanation of the magnitude of the electric field created by a long, straight wire with a positive charge uniformly distributed along its length. A standard textbook account proceeds by modeling the wire as divided into a large number of small segments, each of which acts as a point charge of magnitude $dq$. Each makes a contribution $dE$ to the total field $E$ in accord with the differential form of Coulomb's law:
>
> $$dE = (1/4\pi\varepsilon_0)(dq/s^2)$$
>
> where $s$ is the distance from the charge to an arbitrary point in the field. Integrating over these individual contributions yields the result that the field is at right angles to the wire and that its intensity is given by:
>
> $$E = (1/2\pi\varepsilon_0)(\lambda/r)$$
>
> where $r$ is the perpendicular distance to the wire and $\lambda$ the charge density along the wire. (Woodward and Hitchcock 2003: 3f.)

This example illustrates that the magnitude of an electric field (the explanandum) can be explained by deductively deriving the explanandum from Coulomb's law and initial conditions. In other words, a certain (actual) value of $E$ (the explanandum) is to be expected with certainty conditional on Coulomb's law and actual initial conditions (for instance, the actual charge density $\lambda$ along the wire). Hence, the case at hand is an illustration of the deductive-nomological model of explanation.

However, it is well-known that the classic covering law model of explanation is neither necessary nor sufficient, because it is vulnerable to counter-examples. Although I cannot do justice to the tremendous literature on the covering law model, I will briefly draw attention to three important objections to the covering law approach (cf. Salmon 1989; Craver 2007; Woodward 2009, for surveys to the debate on the covering law model).

First, according to one class of counter-examples, the covering law model allows that the relation between *explanans* and *explanandum* is symmetric. For instance, we can explain why a flagpole with a certain height $h$ casts a shadow with a certain length $l$ by deductively deriving the length $l$ of the shadow from the following premises: the height $h$ of the flagpole, the angle $\alpha$ of the sun above the horizon, and the laws about the rectilinear propagation of light. Moreover, one can also explain the height of the flagpole by deducing the height $h$ from the length $l$ of the shadow, the angle $\alpha$ of the sun above the horizon, and the laws about the rectilinear propagation of light. Intuitively, we are inclined to accept the former case as a genuine explanation. It is hard to accept that the height of the flagpole is explained by the length of the shadow and other facts. To be clear: one needs not deny that one can deduce the height of the flagpole from the lengths of the shadow and other facts. There is nothing wrong with the calculation per se. The intuition of the flagpole scenario is rather that one calculation is explanatory while the other is not.

Second, another counter-example is intended to show that the covering law model implies that some irrelevant factors have to be regarded as explanatory. For instance, suppose that we want to explain why Jeff is not pregnant. In order to explain this fact, we appeal to the true universal general statement 'all men taking birth control pills regularly do not get pregnant'.[1] According to the covering law model, one ought to expect that it is at least highly probable (if not absolutely certain) that a particular man, Jeff, does not get pregnant if the generalization 'all men who take birth control pills regularly do not get pregnant' is true (or well-confirmed) and it is, in fact, the case that Jeff has been taking birth control pills regularly. We ought to expect the explanandum, because 'Jeff does not get pregnant' deductively follows from two premises (by modus ponens): first, 'all men who take birth control pills regularly do not get pregnant'; and, second, 'Jeff is a man who regularly takes birth control pills'. In other words, if one accepts the covering law model, then the generalization 'men taking birth control pills regularly do not get pregnant' and the fact that Jeff takes birth control pills *explains* why Jeff does not get pregnant. Obviously, this is *not* a welcome result for advocates of the covering law approach, because whether men take birth control pills is *entirely irrelevant* for the fact that men do not get pregnant (cf. Woodward 2003: 197–202).

A third counter-example is directed against the inductive-statistical version of the covering law model according to which correlations have to be very strong in order to be explanatory – that is, the conditional probabilities that are expressed in probabilistic generalizations have to be high. According to the so-called Leibniz condition, the conditional probability has to be greater than 0.5 in order to be explanatory. A classic counter-example to this condition is the syphilis-paresis case. Suppose we want to explain why a person suffers from paresis. For all we know, there is only one factor that is significantly correlated with paresis: untreated syphilis. However, the conditional probability of suffering from paresis given that one has already had syphilis that was not treated is 0.1 – that is, P(paresis|syphilis) = 0.1. Moreover, we know that P(paresis|*not*-syphilis) = 0.01. The conditional probability P(paresis|syphilis) is not high and thus violates the requirements of the inductive-statistical model (more precisely, the Leibniz condition). The intuition associated with this case is that P(paresis|syphilis) is neverthe- less *explanatory* because conditionalizing on the fact that a person had untreated syphilis *changes* the probability of the occurrence of paresis.

A common reaction to the trouble of the covering law model stem- ming from these counter-examples is to claim that explanation is about providing *causal information* and not about expectability (cf. e.g. Salmon 1984; Lewis 1986a; Woodward 2003; Craver 2007; Strevens 2009). The basic idea of causal explanation is that appealing to causes neatly refutes the counter-examples introduced above: first, the length of the shadow does not explain the height of the flagpole because the length of the shadow is not a cause of the height of the flagpole. On the other hand, the height of the pole does explain the length of the shadow because the height of the pole causes a shadow of a certain length. Second, although the generalization 'all men taking birth control pills do not get pregnant' is true, it does not do any explanatory work, because tak- ing birth control pills is not a cause of men's failure to get pregnant. Third, the probabilistic statement 'P(paresis|syphilis) = 0.1' is explana- tory because having untreated syphilis is a (positive) probabilistic cause of paresis such that the cause raises the probability of the occurrence of paresis; that is, P(paresis|syphilis)>P(paresis|non-syphilis). An advocate of causal explanation need not demand that causes always raise the probabilities that their effects occur. Causes can also lower these prob- abilities (this phenomenon is captured in my notion of negative proba- bilistic causes in Chapter 8, pp. 214–15). The idea of the paresis-syphilis case is that causes have to alter the probabilities of the effect in order to be explanatory – causes have to make a probabilistic difference. The upshot is that a cause can make a probabilistic difference for the effect, although it might not be the case that we should expect the effect to occur if the cause has occurred.

The common idea of approaches to causal explanation is that to explain an event consists in describing its cause(s). In this section, I will first provide an explication of causal explanation based on my account of causation. Then, I will briefly compare this approach to the influential explication of causal explication by Woodward and Hitchcock.

Causal explanations can be quite naturally expressed in terms of CVS-based notions of causation (see Chapter 8). To explain an actual event *e* is to explain why a proposition of the form '$Y = y$' is actually true (for simplicity's sake, I will begin by assuming that we are mostly interested in the explanation of actual events). Suppose you accept a causal model $M = \langle V, L \rangle$ (for $X, Y \in V$). The occurrence of the actual event *e*, described by the event statement $Y = y$, is causally explained (relative to the accepted model) iff one can provide information about at least one actual cause *c* of *e*, represented by $X = x$. In other words, we explain $Y = y$ by providing information about an actual cause of the explanandum – given that we accept a causal model $M = \langle V, L \rangle$ (for $X, Y \in V$). For instance, take Woodward and Hitchcock's example of the explanation of the magnitude of the electric field (p. 234). According to my account of causal explanations, the actual magnitude of the entire field counterfactually depends on the actual charge density such that if the charge density were different than it actually is, then the magnitude of the entire field would be different than it actually is. Of course, this is a simplified view of an explanation, because more realistic explanations in the special sciences provide information about multiple causes of the explanandum.

It is correct that causal explanation shares certain features with the covering law approach. For instance, causal explanations indirectly rely on (non-universal) laws because these laws are implicit in the causal model. It might also be analogous to the covering law model that it is reasonable to demand that statements about the explanandum and causal model have to be true (I will return to this point shortly). However, the causal approach undermines the foundation of the covering law model: information about the actual charge density is explanatory because it is causally relevant, not because the explanandum was to be expected. Although it might sometimes *coincidentally* be the case that the explanandum is to be expected with a high probability, the epistemic feature of nomic expectability is not a requirement for my account of causal explanation. For this reason, cases of low conditional probabilities can be explanatory: untreated syphilis is a positive probabilistic (i.e. probability-raising) cause of paresis because contracting syphilis makes a counterfactual difference to getting paresis (i.e. if one did not suffer from untreated syphilis, then the probability that one would contract paresis would decrease). That is, untreated syphilis is a positive probabilistic cause of paresis despite the fact that we should *not expect* someone to contract paresis who previously suffered from untreated syphilis. Let me add three remarks and specifications.

First, the explanation of actual events can naturally be expanded to causal *explanation on the type-level* based on the various notions of type-level causation developed in Chapter 8 (such as direct or indirect, deterministic or probabilistic type-level causation). For instance, an event-type $Y$ is causally explained iff one can provide information about a direct deterministic contributing type-level cause $X$ relative to an accepted causal model.

Second, explanation is often taken to depend on the pragmatic context and I agree with this observation. I can account for this pragmatic feature of causal explanation, because the pragmatic aspect of causal explanation is mirrored in the choice of a causal model. Especially, the choice of variables reflects our interest in seeking an explanation and the contrast classes (i.e. the range of the variables in the model) in which we are interested. For instance, in Woodward and Hitchcock's field magnitude case it depends on our explanatory goals whether charge density is represented by a binary variable (such as {having charge density; having no charge density}), or whether the range is more fine-grained and exact – that is, the range is the set of real numbers for quantitative variables (cf. Woodward 2003: 212f., 226–33 on pragmatic aspects of explanations).[2] However, explanations remain to be objective in the following sense: once we have chosen the variables and so on, it is an objective fact about the world whether our pragmatic choice has been wise and the explanation is good.

Third, my account of causal explanation leaves room for *non*-causal explanation. If one is convinced by neo-Russellian arguments (see Chapter 6), then explanation in fundamental physics is non-causal. Moreover, equilibrium explanations and renormalization group explanation might require a non-causal model of scientific explanation (Batterman 2002). Even so, this need not be a threat to models of causal explanation. I do not see why the acausal character of explanation in fundamental physics should be in conflict with my account of causal explanation in the special sciences. However, an account of explanation in physics may exactly look as though it is conceivable that, even in physics, an explanation amounts to describing patterns of counterfactual dependence. In contrast to causal explanation, counterfactual dependence in fundamental physics might have different features – for example, the counterfactuals may not be time-asymmetric because they are grounded in the time-symmetric laws of fundamental physics.

I will now compare my account to another approach to causal explanation. I will focus on the influential approaches by Woodward and Hitchcock. In order to avoid any misunderstandings, I have taken a similar approach to explanation as Woodward and Hitchcock: their model inspired my account. Woodward presents his model of explanations by relying on invariance under interventions:

> Suppose that M is an explanandum consisting in the statement that some variable $Y$ takes a particular value $y$. Then the explanans E for M will consist of (a) a generalization G relating changes in the value(s) of a variable $X$ [...] and changes in $Y$, and (b) a statement (of initial or boundary conditions) that the variable $X$ takes a particular value $x$. A necessary and sufficient explanation for E to be (minimally) explanatory with respect to M is that (i) E and M be true or approximately so; (ii) according to G, $Y$ takes a value $y$ under an intervention in which $X$ takes the value $x$; (iii) there is some intervention that changes the value of $X$ from $x$ to $x'$ where $x \neq x'$, with G correctly describing the value $y'$ that Y would assume under this intervention, where $y' \neq y$. (Woodward 2003: 203)

Similarly, Woodward and Hitchcock (2003) formulate the 'counterfactual approach' to scientific explanations:

> An explanation involves two components, the explanans and the explanandum. The explanandum is a true (or, approximately true) proposition to the effect that some variable (the 'explanandum variable') takes on some particular value. The explanans is a set of propositions, some which specify the actual (approximate) values of variables (explanans variables), and others which specify relationships between the explanans and the explanandum variables. These relationships must satisfy two conditions: they must be true (or approximately so) of the values of the explanans and the explanandum variables, and they must be invariant under interventions. (Woodward and Hitchcock 2003: 6)

Woodward and Hitchcock propose an explication of the concept of a causal explanation in terms of the interventionist theory of causation. The fact to be explained is why a certain event statement of the form $Y = y$ is true. In other words, $Y = y$ is the explanandum. According to Woodward and Hitchcock, $Y = y$ is causally explained iff $Y = y$ can be derived from (a) a set of *true* statements expressing that the explanans variables take their actual values, and (b) a set of generalizations that are invariant under interventions and that specify the relations among $Y$ and the explanans variables. Obviously, this approach is very similar to my own view. In general, I fully agree with Woodward and Hitchcock's emphasis of the fact that explanations consist in exhibiting counter-factual dependence relations, and that generalizations are essential for explanation – however, these generalization may well be non-universal laws, rather than universal laws. Nevertheless, there are at least two important respects in which my account deviates from the interven-tionist account of causal explanation.

First, my account does not rely on the problematic notion of an intervention (see Chapters 3–6). Rather, I propose to understand coun-terfactual dependence in terms of comparative variability semantics (see Chapter 8, pp. 204–9). CVS allows the explanation of events and event-types on which it is physically impossible to intervene (without taking refuge in requiring that interventions be merely logically possible).

Second, I have a concern regarding Woodward and Hitchcock's claim that all of propositions expressing that the explanans variables take certain values have to be *true in the actual world*. My concern is that this claim *contradicts* the interventionist theory of causation, because the explanans *seems* to contain actual *and* merely *hypothetical* information. Let me explain. Suppose, for instance, that an explanation consists in providing information about the actual causes of the dictator's death. Suppose, further, that the actual causal scenario has the structure of a sce-nario of symmetric over-determination; that is, two assassins have shot simultaneously at $t_1$ and they have hit the dictator at $t_2$ (see Chapter 2, p. 66). Woodward was motivated to introduce the notion of a redundancy range by the case of symmetric over-determination, because the evalua-tion of the causal claim 'assassin A's shot was an actual cause of the dic-tator's death' requires that assassin B's shot is held fixed at a non-actual value. This idea is expressed in the interventionist counterfactual 'if A had shot (as A actually did) and *B had not fired* (*contrary to B's actual behaviour*), then the dictator would still have died'. An explanation of the dictator's actual death has to hold fixed B's shot at a non-actual value. This is rep-resented by the statement 'B does not shoot'. However, this statement is false in the actual world (it is not even approximately true). Yet, B's (pos-sible) actions are part of the explanans of the explanation for the dicta-tor's death. Therefore, the requirement that all explanans variables have to take actual values is violated, at least in the case of symmetric over-determination. My own account of causal explanation is not committed to the requirement that all explanans variables have to take their actual values. Rather, the comparative variability theory of causation implies that all of the explanans variables have to take a value in the *redundancy range* with respect to the explanandum variable. As I have indicated earlier (Chapter 2, pp. 37–8, and Chapter 8, pp. 204–5), the redundancy range includes actual and merely possible values of variables.

Third, let me draw the attention to a desideratum of Woodward and Hitchcock's account of explanation. Woodward and Hitchcock repeatedly insist that to explain an event consists in *deriving* the explanandum statement from the laws and other premises (cf., for example, Woodward and Hitchcock 2003: 6, 18–21). They compare this derivation to deductive inference which is used to explicate explanatory relevance in the deductive-nomological model. Yet, they do not provide an account of the validity of inferences from invariant generalizations and, hence, it is not clear by which rules these inferences are legitimized. To be clear, I agree with Woodward and Hitchcock that these inferences within a causal model should not be required to be associated with high subjective probabilities. That is, if we know that the explanandum can be derived from a causal model, we need not expect the occurrence of the explanandum. However, it is an unsatisfactory vagueness and, hence, a desideratum of their approach that they do not provide a logical account that explains how these inferences work. I have to confess that my model of causal explanation does not solve this problem either. To account for rules of correct inferences is also a desideratum for my account, and to work on such an account is a considerably complicated and technical matter that is beyond the scope of this book. However, as already pointed out (Chapter 3, p. 98) in the discussion of the interventionist version of Goodmanian semantics, I am confident that very promising work on inference in causal models, and non-monotonic reasoning in general, has been done by Kvart 1992; Gabbay et al. 1994; Schurz 1997; Pearl 2000; Leitgeb 2004.[3]

## Mechanistic models

In the philosophy of biology, neuroscience, and the social sciences, the concept of a mechanism has been received as an important and fruitful new concept in the past decade. A general and popular claim is that explanations in these sciences (at least, often) describe mechanisms. Among the vast number of explications of the concept of a mechanism, two main definitions stand out. First, Machamer et al.'s definition of a mechanism demarcates the starting-point of the contemporary debate on mechanisms (and mechanistic explanation) in philosophy of science:

> Mechanisms are entities and activities organized such that they are productive of regular changes from start or set-up to finish or termination conditions. (Machamer et al. 2000: 3)

Second, Glennan and Woodward explicate mechanism in terms of an interventionist theory of causation and an invariance theory of laws (cf. also Craver 2007):

> A mechanism for a behavior is a complex system that produces that behavior by the interaction of a number of parts, where the interactions between parts can be characterized by direct, invariant, change-relating generalizations. (Glennan 2002: S344)

> [A] necessary condition for a representation to be an acceptable model of a mechanism is that the representation (i) describe an organized or structured set of parts or components, where (ii) the behavior of each component is described by a generalization that is invariant under interventions, and where (iii) the generalizations governing each component are also independently changeable, and where (iv) the representation allows us to see how, in virtue of (i), (ii) and (iii), the overall output of the mechanism will vary under manipulation of the input to each component and changes in the components themselves. (Woodward 2002: S375)

Although the theories of the nature of a mechanism differ in detail, most mechanists agree that the *description* of a mechanism (a) refers to the state(s) of a complex system, and (b) relies on laws describing the interaction of the parts of a complex system over time. I will not attempt to account for the metaphysics of mechanisms. I will, rather, focus on the characterization of the *representation* of a mechanism: a mechanistic model. In accord with the comparative variability theory, it is a natural view to characterize a mechanistic model for the phenomenon $Y$ in the following way: the mechanism for $Y$ is represented by a causal model $\langle V, L \rangle$ (with $V = \{X_1, ..., X_n, Y\}$) such that:

- the set of variables $\{X_1, ..., X_n\}$ represents the properties of the parts of a complex system $s$;
- the variables $X_1, ..., Xn$ include information about the spatio-temporal organization of the complex system $s$;
- there is at least one directed path leading through some members of $X_1, ..., X_n$ to the outcome (or finish conditions) $Y$; that is, at least one of the $X_i$ has to be the starting point ('the start or set-up conditions', in the words of Machamer et al.) of a directed path to $Y$;[4] and
- each link in a directed path to $Y$ is described by a (non-universal) law (i.e. a member of the set $L$).

I think that this characterization of a causal model representing a mechanism captures three important insights of mechanists: (a) a mechanistic model describes the interactions of the components of a complex system over time, (b) the causal interaction of the components has a causal outcome (thus, we grasp the commonly accepted idea that a mechanism is a mechanism *for* some phenomenon, behaviour or output), (c) a mechanistic model also meets the requirement that the behaviour of the components is represented by non-accidental generalizations. (These generalizations need not be universal deterministic laws. Machamer et al. use the notion of productivity and activity to refer to the non-accidental behaviour of the components of a mechanism, while Glennan, Woodward, and Craver are driven by a similar motivation to employ the notion of invariance.)

Let me point out three advantages for a characterization of mechanistic models based on my comparative variability theory of causation. First, as in the case of causal explanation, my view does not need to rely on interventions in order to characterize the (non-accidental) causal interaction of the components of a complex system. My explication of non-universal laws provides a way to distinguish accidents and laws (see Chapter 5, p. 144). Second, my account of mechanistic models is not committed to accept somewhat obscure entities such as activities. This might be a relief for many empiristically-minded philosophers. However, positing entities such as activities or powers is a metaphysical view that is compatible with my view of mechanistic models. An advocate of activities or powers might nonetheless accept my characterization of mechanistic models. Third, my view of mechanistic models is in agreement with the claim by Woodward, Craver, and Glennan that an *explanatory* mechanistic model $\langle V, L \rangle$, by definition, partly consists of non-accidental generalizations (i.e. non-universal laws, see Chapter 5). It is a consequence of this view that mechanistic explanation relies on generalizations describing the behaviour of the components of a complex system. The commitment to generalizations is a feature that models of mechanistic explanations share with the covering law model.[5] Still, we have to distinguish two claims: it is one thing to say that a model of scientific explanation implies that generalizations are necessary in order to offer an explanation. It is quite another thing to say that generalizations contribute to explanation *because* they give us a reason to *expect* the explanandum to occur. As I have argued, adherents of theories of causal explanation insist that generalizations of the right kind are necessary for providing explanations, although we do not always have reason to believe that the explanandum is likely to occur based on these generalizations.

## Dispositions

The question as to how to understand dispositions is a hot topic in metaphysics of science at the time of writing. Traditionally, dispositions have been analysed as follows:

### Simple conditional analysis
An object x has a disposition D to M iff were x in stimulus conditions S, then x would manifest the behaviour M. (cf. Ryle 1949; Quine 1960; Goodman 1983; as Choi and Fara 2012 point out, this analysis has been implicitly used by many philosophers)

Consider the following canonic example of fragility (the disposition to break) in order to illustrate the simple conditional analysis (SCA): x is fragile iff x were dropped on the ground (i.e. to be dropped is a stimulus condition), then x would break (i.e. x would manifest the behaviour of breaking).

Most philosophers are pessimistic concerning the prospects of SCA because it has been mainly challenged by a type of counter-example involving so-called antidotes (cf. Mumford 1998; Bird 2007). Antidote cases are cases in which x has D and is in stimulus conditions S, yet an interfering factor in the environment of x prevents x from displaying the behaviour M. Consider the canonic example of fragility: a glass is fragile iff the glass were dropped on the ground, then the glass would break (i.e. the glass would manifest the behaviour of breaking). However, the counterfactual is false, if we evaluate it in a world where a glass has been dropped (i.e. stimulus conditions obtain) *and* someone carefully catches the glass before it hits the ground. Catching the glass is an *antidote* to breaking (after the glass has already been dropped). If we accept that antidote worlds are among the closest worlds, then we have a counter-example to SCA.

I think that antidote scenarios are not a serious threat to SCA. My comparative variability semantics (CVS) of counterfactuals provides a straightforward way to deal with antidote-cases: scenarios in which interfering factors (antidotes) occur do not count as a counter-example, because – according to CVS – the counterfactual conditional on the right-hand side of SCA is evaluated in a world in which the object possessing D and undergoing stimulus conditions S is causally isolated with respect to interfering factors. CVS requires that the variables representing interfering factors are held fixed in the redundancy range. For instance, Caroline catches the glass after it was dropped and this is an antidote to its breaking. A binary variable $C$ ranging over the set of possible values {catching; not catching} can represent this antidote. According to CVS, we have to evaluate the counterfactual 'if glass $g$ were dropped, then glass $g$ would break' in a world in which $C$ is held fixed at the possible value *not catching*.[6] Therefore, antidotes are not a counter-example to SCA if the subjunctive conditional of the right-hand side of SCA is evaluated by CVS (in this respect, I side with Gunderson 2002 and Choi 2006, who defend SCA).

A similar reply seems to be appropriate in order to reject counter-examples of so-called fink cases. Fink cases are cases in which an interfering factor leads to a destruction of disposition D (after the stimulus occurred and before the manifestation is realized). For instance, we could imagine a scenario in which a glass is dropped and, before the glass hits the ground, the glass is exposed to extreme heat such that the glass melts. When the melted glass hits the ground it does not break. The intuition behind this finkish scenario is that the extreme increase of the temperature of the glass destroyed the fragility of the glass (i.e. its disposition to break) by changing the molecular structure of the glass. Consequently, the counterfactual on the right-hand side of SCA is false in a fink world. We can draw an analogy to the strategy for dealing with antidote cases: according to CVS, finkish interfering factors (such as exposing a falling glass to extreme heat) do not occur in worlds where the counterfactual of the right-hand side of SCA is evaluated. Thus, finkish scenarios are not a counter-example to SCA, if we use CVS to evaluate the counterfactual associated with the disposition.

The upshot is that CVS provides a straightforward reason to adopt SCA, because one can refute finks and antidotes as counter-examples. In this sense, I have presented a defence of SCA. However, in another sense I fail to meet the original standards of Ryle, Goodman, and Quine. They intended SCA to be a *reductive* analysis of dispositions. Although CVS saves SCA from antidote cases and fink cases, my approach does not satisfy the expectations of a reductive conceptual analysis – provided that reductivity implies no referral to causal notions. It is an analysis that does not avoid the use of causal notions. Whether non-reductive approaches such as the one advocated here are convincing depends on the arguments I presented in Chapter 7.

## Conclusion and outlook

Let me take stock of what I have achieved in this book. The topic is to provide an explication of causation in the special sciences. My project is intended to answer conceptual and semantic questions such as 'What is the concept of causation in the special sciences?' and 'What are the truth conditions of causal claims in these sciences?' I drew a distinction between the semantic project and, on the other hand, metaphysical and methodological inquiries concerning causation. I have been trying to achieve two central goals. The negative goal was to refute a prima facie promising candidate for such an explication: James Woodward's interventionist theory of causation. The positive goal consisted in providing a better explication of causation in the special sciences that could replace the interventionist theory and, at the same time, preserve some of its virtues. I argued for an approach that I labelled the 'comparative variability theory of causation'.

In Part I, I laid a foundation for the discussion about causation in the special sciences. In Chapter 1, I presented criteria of adequacy for an explication of causation in the special sciences. These criteria of adequacy fell into two categories: the naturalist criterion (and its sub-criteria) and the distinction criterion (and its sub-criteria). The naturalist criterion lists (a) features that researchers in the special sciences typically assign to causation, and (b) kinds of causation to which researchers in the special sciences are typically committed. An adequate theory of causation in the special sciences, I claimed, has to explain why causation exhibits these features and to account for these kinds of causation in the form of definitions. In Chapter 2, I reconstructed the influential interventionist theory of causation. I focused on Woodward's interventionist approach according to which $X$ causes $Y$ iff, roughly, there is a possible intervention on $X$ that changes the value of $Y$. I argued that interventionist theories are prima facie (promising to be) in accord with the criteria of adequacy. However, I attempted to show that there are several desiderata that have to be addressed by interventionists. For instance, interventionists ought to account for the explication of a non-universal law, and for the features of time-asymmetry and causal asymmetry in order to meet the naturalist criterion of adequacy.

Part II was intended to achieve the negative goal: that is, to refute the interventionist theory of causation. I raised several objections to Woodward's interventionist theory of causation. I argued that the interventionist theory faces several serious and genuinely internal problems despite its prima facie virtues. The target of my objections was the notion of a possible intervention, which is the key concept of Woodward's approach. In Chapter 3, I diagnosed that interventionists do not provide an account of the truth conditions of interventionist counterfactuals. I argued that this is a desideratum for the interventionist theory. Therefore, I proposed three alternative semantics for interventionist counterfactuals: a possible worlds version and a Goodmanian version of interventionist semantics, and an interventionist version of the suppositional approach. In Chapter 4, I presented several arguments in support of the claim that the interventionists' key notion of a possible intervention is severely problematic, because it leads to a dilemma for interventionists: either interventions are dispensable, or interventions lead to an incorrect evaluation of interventionist counterfactuals. I argued that the dilemma follows from the modal character that Woodward attributes to interventions: according to Woodward, interventions are merely required to be logically possible. I concluded that we should dispense with Woodwardian interventions. In Chapter 5, I addressed a follow-up problem for interventionists: if the notion of a possible intervention is flawed, then this result also affects the account of laws that interventionists have proposed (the invariance theory of laws), because this theory of laws crucially relies on the notion of an intervention. I advocated the following solution: one can account for the no-universal-laws requirement (i.e. one of the sub-criteria of the naturalist criterion) without appealing to the notion of an intervention by defending an alternative explication of non-universal laws in special sciences. In Chapter 6, I focused on the open-systems argument. Interventionists intend this argument to explain (a) why there are no fundamental causal facts and why there are non-fundamental causal facts (the neo-Russellian claim), and (b) why causation has certain features (such as the features of causal asymmetry and time-asymmetry). The open-systems argument essentially depends on a premise that states that it is possible to intervene on open systems (which are taken to be represented by the special sciences). Interventionists believe that the argument provides a positive reason to use the notion of an intervention in order to explicate causation. In response, I argued that the open-systems argument is not sound. So much the worse for interventionists: if the open-systems argument is not sound, then we lack a positive and compelling reason to believe that interventions are indispensable for the enterprise of explicating causation. Moreover, I argued that if the open-systems argument is not sound, interventionists fail to provide (a) a reason to believe the neo-Russellian claim, and (b) an explanation of certain features of causation. Reacting to these problems of the open-systems argument, I presented an alternative argument in favour of the neo-Russellian claim and, in particular, an alternative explanation of the time-asymmetry of causation: the statistical-mechanical approach. Chapter 6 was the only part of the book that explicitly addressed the metaphysics of causation and, thereby, deviated from the main semantic project of explicating causal concepts and providing truth conditions for causal claims. The conclusion I drew from the results of Part II was that we have strong reasons to reject Woodward's interventionist theory of causation.

Part III was an attempt to achieve the positive goal of the book: that is, to replace the troubled interventionist theory of causation with an improved explication of causation in the special sciences. I argued for what I called the comparative variability theory of causation. However, despite the objections to the interventionist theory, my own theory honours the achievements of Woodward's approach. The comparative variability theory preserves several fruitful features of the interventionist theory. In Chapter 7, I addressed the issue that interventionist definitions of causal notions are conceptually circular. However, despite the objection to interventionism raised in Part II, I have defended the conceptually *non-reductive feature* of interventionist definitions. The conceptually non-reductive feature is not unique to interventionist theories of causation: my own theory of causation shares this feature with the interventionist theory (yet, my approach does not rely on interventions). I argued that the comparative variability theory is a better candidate for an adequate explication of causation in the special sciences than the interventionist theory, because my approach is more successful in satisfying the naturalist and the distinction critera of adequacy. Since my approach shares crucial features with the interventionist theory and, yet, eliminates the notion of an intervention, I think that the success of the comparative variability theory supports the dispensability argument against interventions (see Chapter 4, pp. 112–17). I concluded that the comparative variability theory is a more promising candidate for an adequate explication of causation in the special sciences. In Chapter 9, I have highlighted the impact that my theory of causation has on theories of causal explanation, mechanistic models, and the conditional analysis of dispositions.

Let me conclude this chapter by providing an outlook on future research suggested in the light of the results that I have just summarized. Although it is hard to estimate which directions future research will take, I believe that the following topics can be fruitfully connected to the results of my book.

My project has mainly focused on conceptual and semantic questions concerning causation in the special sciences. However, metaphysical questions are forced upon everyone who examines causation also from a philosophical point of view. The core metaphysical questions that have to be answered are mainly: What is the nature of causation? Do causal facts supervene on regularities in the pattern of instantiated intrinsic fundamental properties – as the still dominant Lewisian/Humean view holds (cf. Lewis 1973b, 2004; Earman and Roberts 1999, 2005a, 2005b; Loewer 2008, 2009; Psillos 2009)? Or is causation a modal relation that is not grounded in regularities as Anti-Humeans suggest? Anti-Humean candidates for the truth-makers of causal claims are necessitation relations (Armstrong 1983), dispositions and powers (Cartwright 1989; Hüttemann 2004; Mumford 2004; Bird 2007; Chakravartty 2007), activities (Bogen 2005), subjunctive facts (Lange 2009), primitive facts of temporal evolution (Maudlin 2007), and modal structures (Ladyman and Ross 2007). I think it is an open question which of these metaphysical options best suits my approach (and Woodward's approach). Recall that my approach to causation is based on non-universal law statements. All of the questions that I asked about the metaphysics of causation can equally be asked about non-universal law statements on which Woodward's theory and my approach rely (although we differ on how to explicate generalizations of the special sciences). Are the truth-makers of these lawish statements of the special sciences Lewisian/ Humean regularities? Or is it the case that an adequate and more satisfying understanding of the truth-makers can be found along the lines of an Anti-Humean/Lewisian metaphysics?

Although I relied on the statistical-mechanical account of causation in order to argue for the neo-Russellian claim (see Chapters 6 and 8), the statistical-mechanical account is, at best, an approximation of a story about how causal facts fit into a physical world that seems to be devoid of fundamental causal facts. Although I believe that this account is, indeed, promising and strong, it has to be defended against several objections (cf. Frisch 2007). One has to examine its limitations by showing in greater detail which features of causation it can elucidate and which kinds of causation (e.g. actual and type-level causation) it is able to capture. Neither of these two desiderata of the statistical-mechanical account amounts to a knock-down objection: it is yet undecided whether the statistical-mechanical account can circumvent these obstacles, and the account seems to deserve more attention in the discussion.

The book has touched some, but few, metaphysical issues concerning causation, and I have also neglected methodological issues almost entirely. In Chapter 2, I merely drew a line between Woodward's conceptual project and the debate on causal inference. The latter clearly addresses the methodological question as to how one can derive causal models from statistical data. In Chapter 2, my intention was simply to claim that semantic and methodological questions can be treated independently. I also diagnosed that it would be mistaken to think that Woodward attempts to solve methodological puzzles. For this reason, I restricted my attention to conceptual and semantic questions throughout the book. I stand by my opinion that these are separable projects. However, a *complete* philosophical investigation of causation in the special sciences cannot come to a halt after answering semantic and metaphysical questions. A complete theory of causation also has to improve our understanding of how researchers in these sciences test and confirm causal claims. Hence, what has to be shown is whether my approach is compatible with the methods of causal inference of Pearl (2000) and Spirtes et al. (2000). Although I am not pessimistic regarding the fact that the comparative variability theory can be complemented by these methods of causal inference: it is a fall-back position for me to claim that, if my theory fails to be compatible with these methodological approaches, then it is very likely that Woodward's theory of causation faces similar problems. In general, and apart from the debate on causal modelling, I think it would be fruitful to ask which models of confirmation (e.g. Bayesian and hypothetico-deductive approaches) for causal statements are apt to complement comparative variability theory. These methodological issues give rise to a couple of follow-up questions. The first follow-up question relates to the prospects of formal philosophy in the debate on causation. How are the logical principles of non-monotonic reasoning connected to inferences that are based on causal models containing non-universal laws? How do the non-universal laws (that constitute a causal model) relate to Bayesian probabilities, and to the rules of Bayesian updating of these probabilities?

The second follow-up question concerns the epistemic status of the abstract causal models. My comparative variability theory, as well as Woodward's interventionist theory, defines causal notions *relative to a model*. However, it is not clear how we can decide whether these models are representations of an objective reality that is model-independent. In other words, the question 'Are the models accurate representations of reality because they are, for instance, (approximately) true of or structurally isomorphic to the target system?' is not answered by my theory of causation. Neither is it the case that Woodward's approach addresses the relation between models and reality (cf. Strevens 2007, 2008, for a stimulating discussion of model-relativity). These questions clearly deserve more attention, having launched a longstanding debate on scientific realism and models in science. Nevertheless, I have decided to side-step this debate due to space constraints. Even so, I think that clarifying the relation between (causal) models and 'the world', and explaining under which conditions causal models represent the world, is a major lacuna for, at least, semantic theories of causation such as the interventionist theory and the comparative variability theory. It seems to me that philosophy of science and metaphysics could benefit from connecting the debate on realism and models with the present debate on causation.

# Notes

## 1 Causation in the Special Sciences

1  Cf. Kincaid (2009: s. 3), for a philosophical survey on the features ascribed to causal relations in the social sciences (as ontological and epistemological presuppositions implicit in modelling).
2  Cf. Earman et al. (2002: 297f), and Woodward (2002: 303).
3  However, cf. Schrenk (2007), for a way to amend Lewis's and Armstrong's theories in order to incorporate laws that have 'real exceptions'.
4  Interventionist theories of causation rely on lawish generalizations because, if a causal relation between cause $A$ and effect $B$ obtains, a manipulation of the $A$ leads to a change of $B$ that can be described by an invariant generalization (i.e. an explanatory, not necessarily universal, law-like generalization). Strictly speaking, interventionists insist that generalizations need not be universal. But they also regard this as a challenge: one has to provide a theory of non-universal laws. Cf. Woodward (2003), and Woodward and Hitchcock (2003).
5  Note that Dowe thinks that conserved quantities are only contingently described by the actual conservation laws. In metaphysics of science, such view can be described as categorialism about conserved quantities (cf. Bird 2007: s. 3.1). However, at least in the actual world, laws (and especially conservation laws) seem to matter for the truth conditions of causal statements.

## 2 The Interventionist Explication of Causation

1  Pearl (2000) is an exception, because he pursues the semantic as well as the methodological project with respect to causation.
2  Cf. Kluve (2004: 81f), for a more detailed treatment of these distinctions.
3  Thanks to Siegfried Jaag and Hannes Leitgeb for pointing out the model-relative character of Woodward's theory.
4  Note that Woodward's (2003: 77) original definition AC requires setting all the variables that are not on P on their *actual* values.
5  Other broadly interventionist approaches provide similar, though not strictly equivalent, definitions of actual causation (cf. Hitchcock 2001; Halpern and Pearl 2005; Waters 2007).
6  In the following reconstruction of causal graphs and Bayesian networks, I rely on the standard works by Pearl (2000), Spirtes et al. (2000), Halpern and Pearl (2005: s. 2.1), Jensen and Nielsen (2007: ch. 2) and Williamson (2005: ch. 3).
7  Often, philosophers in this debate talk carelessly about changing an event. Here, a metaphysical problem emerges: the metaphysics of events does *not* allow that a particular event can change and remain the same event. This is apparent if one takes an event to be a property instantiation at a specific point in space-time. 'To change an event $a$' really means to produce another event $b$ that is not identical to $a$. Yet, interventionists are not committed to the strong claim that an intervention changes an event $a$ and that the event $b$ after the

intervention is still identical to *a*. If an actual event is the actual instantiation of an event-type $X = x$, then an intervention on $X$ changes the value of $X$ from $x$ to $x^*$. That is, the result of the intervention is the instantiation of the event-type $X = x^*$ (event *a*) which is *not* identical with the instantiation of the event-type $X = x$ (event *b*).

8  Binary variables are the smallest discrete variables: their range contains only two values.

9  Cf. Meek and Glymour (1994), Scheines (1997), Pearl (2000), Spirtes et al. (2000), and Jensen and Nielsen (2007).

10  For exogenous variable *U*, the probability distribution is $P(U = u|\varnothing)=P(U = u)$ because exogenous variables do not have parents.

11  Cf. especially Pearl (2000: s. 1.4.1), Hitchcock (2001), Halpern and Pearl (2005), and Williamson (2005: 104f. and 122), for detailed treatment of this issue.

12  Pearl (2000: 27) explains this difference of algebraic and structural equations as follows:

> Mathematically, the distinction between structural and algebraic equations is that the latter are characterized by the set of solutions to the entire system of equations, whereas the former are characterized by the solution of each individual equation. The implication is that any subset of structural equations is, in itself, a valid model of reality – one that prevails under interventions. (cf. also Hitchcock 2001: 279–81; and Hoover 2001)

13  For a prominent proponent of this method, cf. Popper (1934: s.12). Williamson (2005: 118–20) presents and objects to this method.

14  Cf. Williamson (2005: ch. 9, ss 9.4–9.7), for a discussion of these constraints including those I present in this section.

15  Cf. Williamson (2005: 125–8), for a presentation of additional alternative methods for discovering causes in the framework of Bayesian networks.

16  Spirtes et al. (2000: 22) point out that *this* formulation is correct, if it is limited to causally sufficient graphs. However, they also provide a revised formulation for graphs that are not causally sufficient (cf. Spirtes et al. 2000: ch. 6).

17  Schurz endorses an even stronger claim: according to him the causal Markov condition is equivalent to the conjunction of the following three claims (which I term 'theorems'). However, I do not wish to commit myself to the truth of Schurz's equivalence claim. Rather, I merely agree with Schurz that the following three claims are theorems that can be derived from the causal Markov condition. Further, I take it that, because these theorems are plausible, they motivate the use of the causal Markov condition. Thanks to Alexander Gepharter for drawing my attention to this point.

18  Cf. Spohn (forthcoming: ch. 14, s. 14.8).

19  Cf. Williamson (2005: 52), for a similar derivation of the causal Markov condition.

20  If the structural equations at hand are indeterministic, then the occurrence of the effect does not necessarily follow. In this case, it can merely be predicted how probable it is that the effect occurs.

21  Lewis held the Whitehead Lectures in March 1999. The lecture manuscript was published for the first time in 2000 in a special issue on causation in the *Journal of Philosophy*, 97: 182–97. The article that I cite was published in 2004 in a volume by Collins et al. (2004a) and is a revised and extended version of Lewis's 2000 paper.

22  Hitchcock (2001), Halpern and Pearl (2005), and Waters (2007) advocate similar interventionist attempts to explicate actual causation.

23  Cf. Woodward (2003: 83); for an elaborated discussion cf. Hitchcock (2001).

24  For brevity's sake, I stick to the assassin example. In the debate, the canonic example for late pre-emption is this one: Billy and Suzy throw rocks at a bottle. Suzy throws first and hits the bottle. Billy throws his rock just a second after Suzy while Suzy's rock is still in the air. But – since the bottle is by that time already smashed to pieces – Billy's rock hits nothing at all at the place where the bottle used to be standing. See Hall (2004: 235). Obviously the structure is the same in my assassin example as Hall's.

25  For a similar argument, see Hitchcock (2001) and Field (2003).

26  Especially the essays in Collins et al. (2004a).

# 3   Counterfactuals: A Problem for Interventionists?

1  See Chapter 2, pp. 27–8, for a more detailed introduction to variables, graphs and causal models. The use of 'variable' in the sense of a random variable in the literature on causation and probability theory is not to be confused with its use in first order predicate logic; that is, in the sense of an individual constant. In the first case, variables refer to properties. In the second case, they refer to individuals. See, for instance, Sinai (1992: 5).

2  Usually the number of values contained in the set is considered to be finite. These values are required to be *mutually exclusive and exhaustive*. The values associated with a variable $X$ are exclusive, if $X$ can only have one value (at a time). They are exhaustive, if the variable has to take one of its possible values. Let me illustrate this with an example, already noted on page 42. Suppose that Mary's income of September 2008 is either high or low; that is, the set of values associated with the income variable $I$ is {high; low}. The set of values is exclusive, because Mary cannot have *both* high income in September 2008 and a low one, too. The set is exhaustive if – as we suppose in the example – Mary's income has to be either high or low, and there is no further value, that it might take, for example 'medium' income.

3  Hiddleston (2005) and Psillos (2007) are among the very few who have diagnosed the problem.

4  One possible interpretation of Woodward's claims is to attribute a verificationist theory of meaning to him. However, I have pointed out that this interpretation conflicts with his declared view, according to which the meaning of causal statements depends on certain mind-independent truth conditions.

5  According to Bennett (2003: 152), Lewisian and Goodmanian views are the main approaches to semantics of counterfactuals. Unfortunately, Bennett's survey skips an important approach: the suppositional theory of counterfactuals.

6  Difficulties with impossible antecedents such as 'If there was a largest natural number ...' or 'If I were Napoleon ...' do not arise (immediately) for someone who is interested in counterfactuals that express causal information.

7  Atomic event statements have the form $X = x$. Complex event statements are built of atomic statements and the logical constants (negation, conjunction, disjunction, material implication). Nothing else is a well-formed formula.

8  Italics are supposed to indicate mentioned expressions in the meta-language.

9  This similarity heuristic applies to *deterministic* cases only. The heuristic has to be modified in order to apply to *indeterministic* cases (cf. Lewis 1986a: 175–84).

10  The general line of critique does not entirely depend on Woodward's particular counter-examples, since Elga (2001), Kment (2006) and Wasserman (2006) present similar counter-examples.

11  Cf., for example, Cartwright (1989), Pearl (2000), Hoover (2001), Frigg and Hartmann (2006).

12  I do not want to be committed to any ontological claim about what possible worlds are. If one would like to be a realist about possible worlds, then one may also say that a specific assignment of values to the variables in a model *represents* a possible world.

13  Kripke (1980: 18f, my italics) emphasizes that even what one calls 'the actual world' is a restricted representation of the world: 'The "actual world" – better the actual state, or history of the world – should not be confused with the enormous scattered object that surrounds us. The latter might be called "the (actual) world", *but it is not the relevant object here.*'

14  Woodward (2003) and Hitchcock (2001) define causal concepts relative to a causal model. Kripke also stresses that small worlds contain only relevant factors (for a certain aim): 'In practice we cannot describe a complete counterfactual course of events *and have no need to do so.* A practical description of the extent to which the "counterfactual situation" differs in the *relevant* way from the actual facts is sufficient; the "counterfactual situation" could be thought of as a miniworld or ministate, *restricted to features of the world relevant to the problem at hand.*' Kripke (1980: 18, emphasis added).

15  Thanks to Daniel Nolan who suggested that this conservative approach is compatible with Stalnaker's semantics. According to Stalnaker (1968), it depends on the pragmatic context how the distance between worlds is measured. It is compatible with this view that interventionists claim that the possibility to intervene measures the distance between worlds in many contexts in which speakers utter causal statements.

16  For simplicity's sake, I presuppose here that the world in which the counterfactual is true or false is the actual world.

17  Italics are supposed to indicate mentioned expressions in the meta-language. As before, Greek letters are used as meta-variables for sentences that express the proposition that a variable take a certain possible value.

18  Cf., e.g., Hitchcock and Woodward (2003: 9) for a commitment to the claim that 'in order to determine whether (and how) the values of $Y$ depend counterfactually on the values of $X$, one must make reference to the causal influence of other variables in specifying what must be held fixed'.

19  Cf. also Mackie (1965: 48f).

20  Pearl (2000: 203) explicitly stresses the point that counterfactuals are evaluated *relative to* a model.

21  Cf. Maudlin (2007: 12f) and Maudlin (2004: 431). I think that the idea of Maudlin's semantics is logically independent of Maudlin's (2007: 17f.) metaphysical thesis about laws: primitivism. This metaphysical view holds that physical laws (of temporal evolution) are ontologically primitive; that is, they do not supervene on anything. Rather, the laws (plus certain facts about property distributions) are the supervenience base for everything else.

22  Let me add three remarks on Maudlin's semantics:

(1) In Step 2, the laws, rather than the values of variables, may be changed. This way, Maudlin's three-step semantics promises to account for *counterlegal* conditionals of the form 'if the laws would be different, then ...' (cf. Maudlin 2007: 31);

(2) In Step 2, the change is supposed to be local: 'Thus "if the bomb dropped on Hiroshima had contained titanium instead of uranium ..." directs us to a moment shortly before the atomic explosion and instructs us to alter the description of the state of the world so that titanium replaces uranium in the bomb. And there is a tacit ceteris paribus condition: leave everything else the same. Don't fool with the position of the plane or the wiring of the bomb or the monetary system of France or anything else' (Maudlin 2007: 34);

(3) In this particular example, Maudlin might be interpreted to appeal to big worlds. However, I think this interpretation is misleading. Maudlin (2007: 18f) argues that possible worlds are model-dependent (or theory-dependent) in the sense that the laws and the variables related by the laws constrain which assignment of values to the variables is consistent.

23  Maudlin (2007: 23) describes this adjustment of the mathematical model in an instructive way: 'The purpose of the antecedent of a counterfactual is to provide instructions on how to locate a Cauchy surface (how to pick a moment in time) and how to generate an altered description of the physical state at that moment. The antecedent is not like an indicative sentence at all; it is more like a command. If the command is carried out a depiction of a situation will result and according to the depiction a certain indicative sentence will be true, but the command is not the same as an indicative sentence.'

24  The second major problem concerns the difficulty in distinguishing between laws and accidentally true general statements.

25  Example: Suppose that the demand (considered as an exogenous variable) does *not stay the same* as in our example. Suppose, instead, that it might increase rapidly. Then the counterfactual (that we evaluated above) is false.

26  Cf. Edgington (1995), (2008) and Leitgeb (2010), for a defence of the suppositional theory.

27  Leitgeb (2010) distinguishes a semantic and a pragmatic meaning of counterfactuals. Based on this distinction, he argues for a complementary combination of a probabilistic variant of Lewis's truth conditional semantics and a suppositional theory along the lines of Adams and Skyrms in order to account for both kinds of meanings.

# 4 Getting Rid of Interventions

1 Alternatively, Woodward's interventionist theory collapses into invariance theories of causation – that is, to the extent to which invariance theories can be formulated without the Woodwardian notion of an intervention.

2 Woodward (2003: 208f.), discusses the constant speed of light, and observes that the generalization 'All physical processes propagate at a speed less than or equal to that of light' is not invariant under physically possible interventions.

3 Of course, one can disagree with the assumption in the antecedent that physical constants are, indeed, causes. However, let us assume that they are causes, at least for the sake of the argument.

4 This is an assumption that Norton is certainly not willing to go along with, because he denies that Newtonian mechanics (and other physical theories after Newton) describe causal relations.

5 Woodward also considers another stronger reading of physical possibility: an intervention $I = i$ is *strongly* physically possible iff $I = i$ is consistent with *actual* initial conditions and the actual laws. However, Woodward – correctly, I think – dismisses strong physical possibility as too demanding a requirement for the interventionist theory of causation (cf. Woodward (2003: 128)).

6 Maudlin (2002: 150) describes the following scenario as 'at least metaphysically possible'. I think that, for this reason, Maudlin's example qualifies for my current purpose.

7 One might object that 'there is a cause of the Big Bang' is conceptually impossible, because, by definition, the Big Bang cannot have a physical cause. Since God need not be a physical being, Maudlin's example neatly avoids the objection that an intervention on the Big Bang is conceptually ill-defined.

8 Although Pearl (2000) relies on the notion of an intervention, a Pearlian intervention differs from a Woodwardian intervention because the former is not committed to the view that interventions have to be modelled as exogenous causes. A Pearlian intervention simply amounts to assigning a specific value to variable $X$ in a causal model while other causal influences are disrupted. This assignment of a value to $X$ is not required to be modelled as the result of an exogenous cause $I$ (i.e. an intervention variable as in the Woodwardian case). I do not raise any objections to the Pearlian kind of interventions.

9 Woodward uses the notion of the redundancy range in order to hold fixed the causal background (cf. Woodward (2003: 83)).

10 In Chapter 6, I present another problem for interventionists: they fail to account for the time-asymmetry of causation.

11 The basic idea of the Ramsey test was historically first used to determine the assertability conditions of indicative conditionals by Adams (1975). Adams analyses the degree of assertability of an indicative conditional in terms of subjective probability: the degree of assertability of 'if $p$, then will $q$' equals the subjective probability of $q$ given $p$. Although this fact is often ignored, Adams (1975: ch. 4) also argues that the suppositional theory can be applied to *counterfactuals* in a slightly modified way. In particular, Skyrms

has developed the most sophisticated account of Adams's original idea. According to Skyrms (1994: 13–15), one determines the pragmatic meaning of a counterfactual in terms of the degree of assertability: the degree of assertability of the counterfactual 'if it were the case that $p$, then it would be that case that $q$' equals the subjectively expected objective conditional probability of $q$ given $p$.

# 5 Non-Universal Laws

1 Cf., for instance, Earman and Roberts (1999); Earman et al. (2002); Lange (2000); Roberts (2004); Woodward (2003), (2007); Maudlin (2007); Strevens (2009).

2 Two terminological clarifications: (1) I will use 'non-universal laws' and '*ceteris paribus* laws' interchangeably; (2) My focus is on *law statements*, rather than on laws themselves – thus, my aim is not to argue for any particular metaphysical claim (such as a regularity view, or a dispositionalist account).

3 Many of the problems I will discuss would be even trickier if one disagreed with Loewer (and others) at this point. Some philosophers (e.g., Cartwright 1983, 1989; Mumford 2004) believe that even fundamental physics deals (at least in part) with non-universal laws. If this were the case, the issue of non-universal laws might turn out to be even more pressing.

4 Lange mistakenly writes 'exponentially'.

5 Cf. Earman et al. (2002: 297f.), Woodward (2002: 303), Kincaid (2004), Roberts (2004). As noted, Cartwright (1983), (1989) and Mumford (2004) dispute the claim that paradigmatic laws of physics conform to the received philosophical picture (e.g. being universal). However, they do not deny that laws in the special sciences are non-universal, have exceptions and so on.

6 Pietroski and Rey (1995: 92) argue that A and C are not '*grue*-like'.

7 Notable exceptions are Mitchell (2000) and Schurz (2002).

8 A useful way to spell out the third dimension of universality could be found in Loewer's use of 'global' (see the quote given on p. 133): 'The dynamical laws of classical mechanics are complete and deterministic. Given the state at any time t they determine the state at any other time. The determination is *global* since the position and momentum of any particle at a time t+r is determined only by the global (i.e. the entire) state of that system at time t. That is, to know how any one particle moves at t+x one has to know something at each particle at t. The dynamical laws and a partial description of state at t (except in special cases) do not entail much about the state of the system at other times and, in particular, don't say much about what any particular particle will (was) doing at t+r' (Loewer 2008: 155). In contrast with the laws of classical mechanics, a special science law (such as the law of supply) is non-global, incomplete, and, thus, seems to provide only a 'partial description' of the phenomenon it describes. Special science laws leave out *other* influences on the phenomenon; that is, circumstances that are not referred to by the law statement itself – as stated in the description of the third dimension of universality (cf. Pietroski and Rey 1995: 89).

9 Note that the concept of a system in Schurz's sense seems to at least coincide with the use of the concept system in the literature on mechanistic explanation in the life sciences – cf. Machamer et al. (2000), Glennan (2002), Bechtel and Abrahamsen (2005), Craver (2007). Mechanists usually conceive a system to be composed of interacting parts. Schurz (2002: s. 5) seems to agree with this characterization of a system when he discusses examples of biological systems.

10 One could object that, even if the restricted reading were the favoured reading, it would not be clear why the corresponding universal statement should be true. Even the universal statement (quantifying only over commodities) is vulnerable to Lange's dilemma and the requirement of relevance. The lesson we should learn from this result is that the responses to these two challenges have to be given with respect to the third and fourth dimension of non-universality.

11 One might want to dispute the claim that even the fundamental laws do not apply to *everything* – contra Schurz (2002), Hüttemann (2007). One might object that the fundamental laws, for instance, do not apply to angels and numbers. However, I think that, even if this were the case, we could preserve the universality$_2$ for the fundamental laws by exactly the same strategy that I just used for preserving universality$_2$ for lawish statements in the special sciences. Further, my arguments do not have to rely on the characterization of fundamental physical laws that Schurz and Hüttemann provide.

12 A typical example is provided by causal models in econometrics: according to these models, the causal influence of a variable in isolation is described by a single structural equation. Each one of those single equations might be called 'inertial law'. However, the whole causal *model* (i.e. a set of equations) provides an overall output resulting from the interaction of various causal factors – cf. Cartwright 1989: ch. 4, section 5; Pearl 2000: ch. 5.

13 Thanks to Tim Maudlin and Michael Strevens for suggesting this amendment.

14 Similar approaches such as Lange's (2000: 103, 2009: 29) and Mitchell's (2000) stability theories as well as Ladyman and Ross's (2007) 'real pattern'-approach might also work for my argument.

15 One could, of course, say that assuming that other causes are absent is a limiting case of holding these factors fixed.

16 My approach does not differ from Hitchcock and Woodward's invariance theory concerning the *non-reductive* feature of the explication of the concept of a lawish generalization. Both explications are non-reductive, because they use causal and nomological concepts in the explicans. I agree with Hitchcock and Woodward that the non-reductive character of an explication is unproblematic, as long as the explication is not viciously circular. For a more detailed defence of non-reductive explication see Chapter 7 and cf. Woodward (2003: 103f., 2008: 203f.), Strevens (2008: 186).

17 Invariance theories usually refer to counterfactual situations in which the factors governed are undisturbed (i.e. the counterfactual situation involves elements of idealization and abstraction) in order to state the truth-conditions of law statements. A *methodological* question immediately comes to mind: how can one *test* these statements? How do scientists actually test statements of this kind in practice? Various authors have addressed these

questions (cf. Kincaid (1996), Cartwright (2007), Steel (2007), Reiss (2008), Hüttemann (2012).

18  Cf. Woodward (2003, 2008) and Woodward and Hitchcock (2003), for a more detailed account of this invariantist strategy to distinguish lawish statements from accidentally true generalizations. Reutlinger et al. (2011: s. 6) provide an easily accessible survey to invariance theories.

# 6  Woodward Meets Russell: Does Causation Fit into the World of Physics?

1  Russell's denial of causation in physics was supported by many philosophers of physics and philosopher-physicists of his day such as Wien (1915: 246) and Mach (1900: 278), while philosophers such as Toulmin, Havas, and Hund argued in favour of a Russellian view of physics in mid-twentieth century, cf. Scheibe (2006: 218–23), for a historical survey.

2  Another important reason to believe that causes are not sufficient for the occurrence of their effects consisted in the acknowledgement of probabilistic causation.

3  Cf. Feynman (1965: 108–12), who employs the law of gravitation as an example for the time-symmetry of physical laws, and Albert (2000: 2–9), who illustrates the feature of time-symmetry with Newton's laws of motion. David Albert (2000: 9–21) also defends the claim that the fundamental dynamical laws are time-symmetric. Yet, Albert also maintains that the meaning of time-asymmetry has to be adjusted in order to fit physical theories after Newtonian physics (e.g. classical electrodynamics, statistical mechanics, special and general relativity theory, quantum mechanics, quantum field theory, string theory). The upshot of Albert's discussion is that a modified sense of time-symmetry holds also for physics after Newton. North (2008) is a recent discussion of different readings of time-symmetry in physics.

4  Cf. Woodward (2007: 66), for a similar presentation of the problem.

5  An alternative (and more realistic) stance could be: fundamental physical laws often describe local systems as isolated. These local physical systems are only a small portion of the entire universe. Whether one can influence an isolated system from the environment of the system depends on the *kind* of isolation we have in mind. That is, we have to decide whether the system is isolated from actually occurring, physically possible or logically possible interfering factors.

6  However, Woodward's worlds should be understood as model worlds or small worlds; that is, assignments of values to variables in a causal model (cf. Pearl 2000: 207). In this respect, Woodwardian worlds differ from Lewisian worlds because the latter are typically as detailed as the real spatio-temporal entity we inhabit (cf. Hüttemann 2004: 113).

7  There are other optional semantics for counterfactuals such as various Goodmanian and suppositionalist approaches (see Chapter 3).

8  Although this is a possible rejoinder, I doubt that most interventionists would be comfortable endorsing claims about the essences of things.

9  This argument is also sound if we do not assume that fundamental laws describe the whole universe and, instead, suppose that the fundamental

laws of physics describe (local) isolated systems such that being isolated is *essential* for being the system under description.

10  Note that Elga carefully observes that non-$c$ is a time-reversed state (cf. Elga 2001: 316). I will omit the details of this point here.

11  Cf. Ney (2009: 753), for a more down-to-earth billiard ball example illustrating the same point.

12  As I have remarked earlier, Woodward refers to Strevens's (2003) account of objective probabilities. However, the analogy with the account by Albert, Kutach, and Loewer is even more striking: they intend to account for the truth-makers of causal statements in a similar way as Strevens approaches the truth-makers of probability statements. However, it might be worth exploring the unique features of Strevens's account more deeply at some future point.

13  Thanks to Matt Farr and Jacob Rosenthal who became concerned at this point when discussing an earlier draft of this chapter.

# 7  In Defence of Conceptually Non-Reductive Explications of Causation

1  Similar accounts are presented by Hausman (1998), Hitchcock (2001), Halpern and Pearl (2005).

2  According to Lewis (1973a: 24–6), a counterfactual is *vacuously true* iff its antecedent expresses an impossible proposition. This might be a controversial claim. The case of vacuous truth need not concern us here, because causal claims have to express contingent propositions.

3  This is, of course, a charitable reading of Lewis's semantics. One might claim that Lewis's similarity measure *implicitly* refers to causal facts and causal intuitions. Woodward (2003: 133–45) seems to suggest an objection along these lines.

4  Cf. Collins et al. (2004b), Elga (2001), Hitchcock (2001), Woodward (2003), Schaffer (2004b), Kment (2006), Wasserman (2006), and Price and Corry (2007), for a detailed list of counter-examples to Lewis's counterfactual theory of causation.

5  Most papers in Collins et al. (2004a) are attempts to do precisely this.

6  Another reason to adopt the reductivity criterion is based on Russellian metaphysical concerns about causation on the fundamental physical level. Yet, as I argued, in Chapter 6, conceptually non-reductive theories of causation are compatible with a metaphysical claim that neo-Russellians hold: there are no fundamental causal facts, but there are non-fundamental causal facts of the special sciences.

7  Notice that Lewis deals with actual causation; that is, the relata of a causal relation are events. Cartwright and Woodward focus on type-level causation: that is, the causal relata are random variables. Nevertheless, their theories can also account for actual causation – see Woodward (2003: 74–85); Halpern and Pearl (2005: 583). This difference is a minor point here, because the question as to whether one is tied to a reductive methodology applies to the analysis of actual as well as of type-level causation.

8  Cartwright claims that the general problem behind this example is Simpson's Paradox; that is, 'any association [...] between two variables which holds in

a given population can be reversed in the sub-populations by finding a third variable which is correlated with both' (Cartwright 1983: 24). Cf. Cartwright (2007: 64), for a more recent statement of the same problem with respect to Simpson's paradox.

9  Cf. Gupta (2008: ss 2.5–2.7), for a detailed survey on explicit and implicit definitions. A famous example of an implicit definition is Lewis's definition of objective probability (cf. Lewis 1994).

10  Hitchcock (2006: 450) also indicates that Jackson's theoretical role view is a proper methodological reflection of non-reductive analysis.

11  Cf. Collins et al. (2004b), for a detailed list of counter-examples (most importantly various scenarios of pre-emption and probabilistic causation). Cf. Cartwright (1983), Hitchcock (2001), Woodward (2003), and Halpern and Pearl (2005), for non-reductive accounts that propose convincing treatments for those counter-examples to Lewisian counterfactual theories of causation.

12  In the sciences, one finds very restricted abstract and/or idealized models of phenomena including the paradigmatic causal claims listed in Chapter 1 (p. 7), and the causal models presented in Chapter 2 (pp. 39–51) (cf. Cartwright 1989; Frigg and Hartmann 2006; Frigg forthcoming).

# 8  The Comparative Variability Theory of Causation

1  It is a distinct question how these models can be tested and why we should believe that these models are good representations of economic reality. However, as I have indicated earlier, I will not attempt to answer these methodological questions here.

2  Although I agree with Hitchcock on the importance of counterfactual dependence and the assumption of holding other causes fixed, I depart from Hitchcock's path because he believes that facts about interventions fix the meaning of causal statements and of non-universal laws (structural equations).

3  King et al. (1994) and Morgan and Winship (2007) present similar 'potential outcome' approaches in sociology.

4  Again, italics are supposed to indicate meta-language. As before, Greek letters are used as meta-variables for sentences that express the proposition that a variable take a certain possible value.

5  Notice that Woodward's (2003: 77) original definition AC requires setting all the variables that are not on P on their actual values.

6  On the same page, Field goes on to explain in fn. 26: 'When C, E, and A are true, saying that E counterfactually depends on C, *holding A fixed*, is simply to say that *if A and not-C* were the case then not-E would be the case. This is an ordinary non-backtracking counterfactual, though one whose *antecedent involves multiple locations*' (Field 2003: 452, emphasis added).

7  Even the SM-account presented in Chapter 6 is, in principle, open to Humean and non-Humean readings.

# 9  Consequences

1  Although I will not discuss this case, the example also works for statistical generalizations expressing strong correlations.

2   The pragmatic aspect only disappears if we assume that there is a causal model with a set of 'natural' variables **N** such that the event-types that are associated with the variables in **N** – for example, $X = x$ – refer to fundamental natural properties.

3   For the reason that Woodward and Hitchcock compare their approach to David Lewis's theory of causal explanations, I will briefly discuss it. Lewis advocates a theory of causal explanations according to which 'to explain an event is to provide some information about its causal history' (Lewis 1986a: 217). Causation is in this case understood in terms of Lewis's counterfactual theory of causation. Explanation is a matter of degree in the sense that an explanation can be more or less complete depending on how much of the information of the explanandum's causal history one supplies (cf. Lewis 1986a: 238). Lewis claims that causal explanations 'often provide general explanatory information about events of a given kind' (cf. Lewis 1986a: 225). I have no reason to deny that causal explanations can be on the type-level. Yet, Lewis merely provides a theory of *actual* causation, and it is, at least, unclear how Lewis explicates type-level causal notions and type-level causal claims such as 'type-C events cause type-E events'. Hence, Lewis fails to offer a theory of type-level causal explanation. By contrast, Woodward's interventionist approach and my comparative variability theory of causation offer explications of type-level causal claims and a model of type-level causal explanation (cf. also Strevens (2009) for a theory of causal explanation of events, event-types, and laws).

4   The simplest causal model of a mechanism, hence, is a model according to which there is a single directed causal path from, say, $X_1$ via $X_2$ ... $X_n$ to $Y$.

5   This is correct unless one is a singularist concerning causation. Woodward, Glennan, Craver, and I adopt generalist views of causation.

6   A remark on antidotes: as I argue in Chapter 5, laws of deviation describe the behaviour of the glass caused by an antidote.

# Bibliography

Adams, Ernest (1975) *A Theory of Conditionals. An Application of Probability to Deductive Logic* (Dordrecht: Reidel).

Albert, David (2000) *Time and Chance* (Cambridge, MA: Harvard University Press).

Armstrong, David (1983) *What Is a Law of Nature?* (Cambridge, MA: Cambridge University Press).

Bartelborth, Thomas (2007) *Erklären* (Berlin/New York: de Gruyter).

Batterman, Robert (2002) *The Devil in the Details* (Oxford: Oxford University Press).

Baumgartner, Michael (2008) 'Regularity Theories Reassessed', *Philosophia*, 36, 327–54.

Beatty, John (1995) 'The Evolutionary Contingency Thesis', in G. Wolters and G.J. Lennox (eds), *Concepts, Theories, and Rationality in the Biological Sciences* (Pittsburgh: Pittsburgh University Press): 45–81.

Bechtel, William and Adele Abrahamsen (2005) 'Explanation: A Mechanist Alternative', *Studies in History and Philosophy of Biological and Biomedical Sciences*, 36: 421–41.

Beebee, Helen (2007) 'Hume on Causation: The Projectivist Interpretation', Huw Price and Richard Corry (eds), *Causation, Physics, and the Constitution of Reality. Russell's Republic Revisited* (Oxford: Clarendon Press): 224–9.

Beebee, Helen, Christopher Hitchcock and Peter Menzies (eds) (2009) *The Oxford Handbook of Causation* (Oxford: Oxford University Press).

Beed, Clive and Cara Beed (2000) 'Is the Case for Social Science Laws Strengthening?', *Journal for the Theory of Social Behaviour*, 30(2): 131–53.

Bennett, Jonathan (2003) *A Philosophical Guide to Conditionals* (Oxford: Oxford University Press).

Bird, Alexander (2007) *Nature's Metaphysics: Laws and Properties* (Oxford: Oxford University Press).

Bogen, James (2005) 'Regularities and Causality; Generalizations and Causal Explanations', in C. Craver and L. Darden, 'Mechanisms in Biology', *Studies in History and Philosophy of Biological and Biomedical Sciences*, 36: 397–420.

Boghossian, Paul (1996) 'Analyticity Reconsidered', *Nous*, 30(3): 360–91.

Boghossian, Paul (2000) 'Knowledge of Logic', in Paul Boghossian and Christopher Peacock (eds), *New Essays on the A Priori* (Oxford: Clarendon Press): 229–54.

Braithwaite, R.B. (1959) *Scientific Explanation* (Cambridge: Cambridge University Press).

Burgess, J. (2008) 'When is Circularity in Definitions Benign?', *Philosophical Quarterly*, 58: 214–233.

Callender, Craig (2011) 'Thermodynamic Asymmetry in Time', in Edward N. Zolta (ed.), *The Stanford Encyclopedia of Philosophy*, fall 2011, available at http://plato.stanford.edu/archives/fall2011/entries/time-thermo.

Callender, Craig and Jonathan Cohen (2010) 'Special Sciences, Conspiracy and the Better Best System Account of Lawhood.', *Erkenntnis*, 73(3): 427–47.

Campbell, John (2007) 'An Interventionist Approach to Causation in Psychology', in Alison Gopnik and Laura Schulz (eds), *Causal Learning: Psychology, Philosophy and Computation* (Oxford: Oxford University Press): 58–66.

Carnap, Rudolf (1950) *Logical Foundations of Probability* (Chicago: University of Chicago Press).

Carnap, Rudolf (1960) *Meaning and Necessity. A Study in Semantics and Modal Logic* (Chicago: University of Chicago Press).

Carnap, Rudolf (1966) *An Introduction to the Philosophy of Science* (New York: Dover Publications).

Cartwright, Nancy (1983) *How the Laws of Physics Lie* (Oxford: Oxford University Press).

Cartwright, Nancy (1989) *Nature's Capacities and their Measurement* (Oxford: Oxford University Press).

Cartwright, Nancy (1999) *The Dappled World* (Cambridge: Cambridge University Press).

Cartwright, Nancy (2007) *Hunting Causes and Using them. Approaches in Philosophy and Economics* (Cambridge: Cambridge University Press).

Cartwright, Nancy (2009) 'Causality, Invariance, and Policy', in H. Kincaid and D. Ross (eds), *The Oxford Handbook of Economics* (Oxford: Oxford University Press): 410–23.

Chakravartty, Anjan (2007) *A Metaphysics for Scientific Realism. Knowing the Unobservable* (Cambridge: Cambridge University Press).

Chalmers, David (1996) *The Conscious Mind. In Search of a Fundamental Theory* (New York: Oxford University Press).

Choi, Sungho (2006) 'The Simple vs. Reformed Conditional Analysis of Dispositions', *Synthese*, 148: 369–7.

Choi, Sungho and Michael Fara (2012) 'Dispositions', in Edward N. Zalta (ed.) *The Stanford Encyclopedia of Philosophy*, spring 2012 edn, available at http://plato.stanfordedu/archives/spr2012/entries/dispositions/.

Collins, John, Ned Hall and L.A. Paul (eds) (2004a) *Causation and Counterfactuals* (Cambridge, MA: MIT Press).

Collins, John, Ned Hall and Laurie Paul (2004b) 'Counterfactuals and Causation. History, Problems and Prospects', in J. Collins, N. Hall and L. Paul, *Causation and Counterfactuals* (Cambridge, MA: MIT Press): 1–58.

Cook, T. and D. Campbell (1979) *Quasi-Experimentation* (Boston: Houghton Mifflin).

Craver, Carl (2007) *Explaining the Brain. Mechanisms and the Mosaic Unity of Neuroscience* (Oxford: Clarendon Press).

Dowe, Phil (2000) *Physical Causation* (Cambridge, MA: Cambridge University Press).

Eagle, Antony (2007) 'Pragmatic Causation', in Huw Price and Richard Corry (eds), *Causation, Physics, and the Constitution of Reality. Russell's Republic Revisited*, (Oxford: Clarendon Press): 156–90.

Eagle, Antony (2011) *Philosophy of Probability. Modern and Contemporary Readings* (London: Routledge).

Earman, John and John Roberts (1999) 'Ceteris Paribus: There is No Problem of Provisos', *Synthese*, 118: 439–78.

Earman, John and John Roberts (2005a) 'Contact with the Nomic: A Challenge for Denizens of Human Superverience (Part One)', *Philosophical and Phenomenological* Research 71:1–22.

Earman, John and John Roberts (2005b) 'Contact with the Nomic: A challenge for Denizens of Human Superverience (Part Two) *Philosophical and Phenomenological,* 71:253–86.

Earman, J., J. Roberts and S. Smith (2002) 'Ceteris Paribus Lost', *Erkenntnis,* 57(3), Special Issue: 281–301.

Edgington, Dorothy (1995) 'On Conditionals', *Mind,* 104: 235–329.

Edgington, Dorothy (2008) 'Counterfactuals', *Proceedings of the Aristotelian Society,* 108: 1–21.

Elga, Adam (2001) 'Statistical Mechanics and the Asymmetry of Counterfactual Dependence', *Philosophy of Science,* 68: S313–24.

Engle, Robert, David Hendry and Jean-Francois Richard (1983) 'Exogeneity', *Econometrica,* 51: 277–304.

Feynman, Richard (1965) *The Character of Physical Law* (London: Penguin).

Field, Hartry (1980) *Science Without Numbers: A Defence of Nominalism* (Oxford: Blackwell).

Field, Hartry (2003) 'Causation in a Physical World', in M. Loux and D. Zimmerman (eds), *The Oxford Handbook of Metaphysics* (Oxford: Oxford University Press): 435–60.

Freedman, David (1997) 'From Association to Causation via Regression', in V. McKim and S. Turner (eds), *Causality in Crisis? Statistical Methods and the Search for Causal Knowledge in the Social Sciences* (Notre Dame: University of Notre Dame Press): 113–61.

Frigg, Roman (forthcoming) *Models and Theories* (Durham: Acumen).

Frigg, Roman, and Stephan Hartmann (2006) 'Models in Science', Edward N. Zalta (ed.), *Stanford Encyclopedia of Philosophy,* summer 2009 edn, available at http://plato.stanford.edu/archives/sum2009/entries/models-science/.

Frisch, Mathias (2007) 'Causation, Counterfactuals, and Entropy', in Huw Price and Richard Corry (eds), *Causation, Physics, and the Constitution of Reality. Russell's Republic Revisited* (Oxford: Clarendon Press): 351–95.

Frisch, Mathias (2009a) '"The Most Sacred Tenet"? Causal Reasoning in Physics', *British Journal for Philosophy of Science,* 60: 459–74.

Frisch, Mathias (2009b) 'Causality and Dispersion: A Reply to John Norton', *British Journal for Philosophy of Science,* 60: 487–95.

Frisch, Mathias (2010) 'Kausalität in der Physik', available at http://www.philosophy.umd.edu/Faculty/mfrisch/papers/kausalitaet.pdf.

Gabbay, Dov M., C.J. Hogger and J.A. Robinson (eds) (1994) *Handbook of Logic in Artificial Intelligence and Logic Programming, Volume 3: Nonmonotonic Reasoning and Uncertain Reasoning* (Oxford: Clarendon Press).

Gadenne, Volker (2004) *Philosophie der Psychologie* (Bern: Verlag Hans Huber).

Gillies, Donald (2001) 'Critical Notice. Judea Pearl "Causality"', *British Journal for Philosophy of Science,* 52: 613–22.

Glennan, Stuart (2002) 'Rethinking Mechanist Explanation', *Philosophy of Science,* 69 (supplement): S342–53.

Glock, Hans-Johann (2003) *Quine and Davidson on Language, Thought and Reality* (Cambridge: Cambridge University Press).

Glymour, Clark (2010) 'What Is Right with "Bayes Net Methods" and What Is Wrong with "Hunting Causes and Using Them"?', *British Journal for Philosophy of Science,* 61(1): 161–211.

Goldthorpe, John (2001) 'Causation, Statistics, and Sociology', *European Sociological Review,* 17: 1–20.

Goodman, Nelson (1983) *Fact, Fiction and Forecast*, 4th edn (Cambridge, MA: Cambridge University Press).

Granger, Clive (1962) 'Economic Processes Involving Feedback', *Information and Control*, 6: 28–48.

Granger, Clive (1969) 'Testing for Causality and Feedback', *Econometrica*, 37: 424–38.

Granger, Clive (2007) 'Causality in Economics', in Peter Machamer and Gereon Wolters (eds), *Thinking about Causes* (Pittsburgh: Pittsburgh University Press): 284–96.

Gundersen, Lars (2002) 'In Defence of the Conditional Account of Dispositions', *Synthese*, 130: 389–411.

Gupta, Anil (2008) 'Definitions', Edward Zalta (ed.), *Stanford Encyclopedia of Philosophy*, spring 2009 edn, available at http://plato.stanford.edu/archives/spr2009/entries/definitions/.

Hale, Bob and Crispin Wright (2000) 'Implicit Definition and the A Priori', in Paul Boghossian and Christopher Peacock (eds), *New Essays on the A Priori* (Oxford: Clarendon Press): 286–319.

Hall, Ned (2004) 'Two Concepts of Causation', in J. Collins, N. Hall and L. Paul (eds) *Causation and Counterfactuals* (Cambridge, MA: MIT Press): 181–204.

Halpern, Joseph and Judea Pearl (2005) 'Causes and Explanations: A Structural-Model Approach. Part I: Causes', *British Journal for the Philosophy of Science*, 56: 843–87.

Hausman, Daniel (1988) 'Ceteris Paribus Clauses and Causality in Economics', *PSA Proceedings 1988. Volume Two: Symposia and Invited Papers*: 308–16.

Hausman, Daniel (1992) *The Separate and Inexact Science of Economics* (Cambridge, MA: Cambridge University Press).

Hausman, Daniel (1998) *Causal Asymmetries* (Cambridge: Cambridge University Press).

Hausman, Daniel (2009) 'Laws, Causation, and Economic Methodology', in H. Kincaid and D. Ross (eds), *The Oxford Handbook of Economics* (Oxford: Oxford University Press): 35–54.

Hausman, Daniel and James Woodward (1999) 'Independence, Invariance, and the Causal Markov Condition', *British Journal for the Philosophy of Science*, 50: 1–63.

Hausman, Daniel and James Woodward (2004) 'Modularity and the Causal Markov Condition: A Restatement', *British Journal for the Philosophy of Science*, 55: 147–61.

Heath, Joseph (2005) 'Methodological Individualism', Edward N. Zalta (ed.), *Stanford Encyclopedia of Philosophy* (spring 2011 edn), available at http://plato.stanford.edu/archives/spr2011/entries/methodological-individualism/.

Heckman, James (2008) 'Econometric Causality', *International Statistical Review*, 76(1): 1–27.

Hedström, Peter and Ylikoski, Petri (2010) 'Causal Mechanisms in the Social Sciences', *Annual Review of Sociology*, 36: 49–67.

Hempel, Carl (1965) *Aspects of Scientific Explanation and Other Essays in the Philosophy of Science* (New York: The Free Press).

Hendry, David (2004) 'Causality and Exogeneity in Non-stationary Economic Time-Series Causality', Metaphysics and Methods Technical Report, CTR 18-04, Center for Philosophy of Natural and Social Science, London School of Economics.

Hicks, John (1979) *Causality in Economics* (New York: Basic Books).

Hiddleston, Eric (2005) 'Review of Making Things Happen by James Woodward', *Philosophical Review*, 114(4): 545–7.

Hitchcock, Christopher (2001) 'The Intransitivity of Causation Revealed in Equations and Graphs', *Journal of Philosophy*, 98: 273–99.

Hitchcock, Christopher (2006) 'Conceptual Analysis Naturalized: A Methodological Case Study', *Journal of Philosophy*, 103(9): 427–51.

Hitchcock, Christopher (2007) 'What Russell Got Right', in Huw Price and Richard Corry (eds), *Causation, Physics, and the Constitution of Reality. Russell's Republic Revisited* (Oxford: Clarendon Press): 45–65.

Hitchcock, Christopher (2010) 'Probabilistic Causation', Edward N. Zalta (ed.), *Stanford Encyclopedia of Philosophy* (autumn 2010 edn), available at http://plato.stanford.edu/archives/fall2010/entries/causation-probabilistic/.

Hitchcock, C. and Woodward, J. (2003) 'Exploratory Generalization Part II: Plumbing Exploratory Depth', *Nôus*, 37: 181–99.

Hoover, Kevin (1988) *New Classical Macroeconomics* (Oxford: Basil Blackwell).

Hoover, Kevin (2001) *Causality in Macroeconomics* (Cambridge: Cambridge University Press).

Humberstone, Loyd (1997) 'Two Types of Circularity', *Philosophy and Phenomenological Research*, 58: 249–80.

Hüttemann, Andreas (2004) *What's Wrong With Micro-Physicalism?* (London: Routledge).

Hüttemann, Andreas (2007) 'Naturgesetze', in Andreas Bartels and Manfred Stöckler (eds), *Wissenschaftstheorie* (Paderborn: mentis): 135–53.

Hüttemann, Andreas (2012) 'Ceteris Paribus Gesetze in der Physik', in Michael Esfeld (ed.), *Philosophie der Physik* (Berlin: Surkamp Verlage): 390–410.

Jackson, Frank (1998) *From Metaphysics to Ethics. A Defense of Conceptual Analysis* (Oxford: Clarendon Press).

Jensen, Finn and Thomas Nielsen (2007) *Bayesian Networks and Decision Graphs* (New York: Springer).

Joyce, James (1999) *The Foundations of Causal Decision Theory* (Cambridge: Cambridge University Press).

Kincaid, Harold (1996) *Foundations of the Social Sciences* (Cambridge: Cambridge University Press).

Kincaid, Harold (2004) 'Are There Laws in the Social Sciences? Yes', in Christopher Hitchcock (ed.), *Contemporary Debates in the Philosophy of Science* (Oxford: Blackwell): 168–87.

Kincaid, Harold (2009) 'Explaining Growth', in H. Kincaid and D. Ross (eds), *Oxford Handbook of Economics* (Oxford: Oxford University Press): 455–75.

Kincaid, Harold and Don Ross (eds) (2009) *Oxford Handbook of Economics* (Oxford: Oxford University Press).

King, Gary, Robert O. Keohane and Sidney Verba (1994) *Designing Social Inquiry. Scientific Inference in Qualitative Research* (Princeton: Princeton University Press).

Klein, David (1987) 'Causation in Sociology Today: A Revised View', *Sociological Theory*, 5(1): 19–26.

Kluve, Jochen (2004) 'On the Role of Counterfactuals in Inferring Causal Effects', *Foundations of Science*, 9: 65–101.

Kment, Boris (2006) 'Counterfactuals and Explanation', *Mind*, 115: 261–309.

Kripke, Saul (1980) *Naming and Necessity* (Cambridge, MA: Harvard University Press).

Kutach, Douglas (2007) 'The Physical Foundation of Causation', in Huw Price and Richard Corry (eds), *Causation, Physics, and the Constitution of Reality. Russell's Republic Revisited* (Oxford: Clarendon Press): 327–50.

Kvart, Igal (1992) 'Counterfactuals', *Erkenntnis*, 36: 1–41.

Kvart, Igal (2001) 'The Counterfactual Analysis of Cause', *Synthese*, 127: 389–427.

Ladyman, James and Don Ross (2007) *Every Thing Must Go. Metaphysics Naturalized* (Oxford: Oxford University Press).

Lange, Marc (1993) 'Natural Laws and the Problem of Provisos', *Erkenntnis*, 38: 233–48.

Lange, Marc (2000) *Natural Laws in Scientific Practice* (Oxford: Oxford University Press).

Lange, Marc (2002) 'Who's Afraid of Ceteris Paribus Laws? Or: How I Learned to Stop Worrying and Love Them', *Erkenntnis*, 57(3): 407–23.

Lange, Marc (2009) *Laws and Lawmakers* (Oxford: Oxford University Press).

Leitgeb, Hannes (2004) *Inference on the Low Level. An Investigation into Deduction, Nonmonotonic Reasoning, and the Philosophy of Cognition* (Berlin/New York: Springer).

Leitgeb, Hannes (2012) 'A Probabilistic Semantics for Counterfactuals. Part A', *The Review of Symbolic Logic*, 5: 26–36.

Leuridan, Bert (2010) 'Can Mechanisms Really Replace Laws of Nature?', *Philosophy of Science*, 77: 37–340.

Lewis, David (1973a) *Counterfactuals* (Oxford: Blackwell).

Lewis, David (1973b) 'Causation', in David Lewis, *Philosophical Papers II* (Oxford: Oxford University Press): 159–72.

Lewis, David (1979) 'Counterfactual Dependence and Time's Arrow', in David Lewis (1986), *Philosophical Papers II* (Oxford: Oxford University Press): 32–51.

Lewis, David (1986a) *Philosophical Papers II* (Oxford: Oxford University Press).

Lewis, David (1986b) 'Causal Explanation', in David Lewis, *Philosophical Papers II* (Oxford: Oxford University Press): 214–40.

Lewis, David (1994) 'Humean Supervenience Debugged', *Papers in Metaphysics and Epistemology* (Cambridge: Cambridge University Press): 224–47.

Lewis, David (2004) 'Causation as Influence', in John Collins, Ned Hall and L.A. Paul (eds), *Causation and Counterfactuals* (Cambridge, MA: MIT Press): 75–106.

Liebesman, David (2011) 'Causation and the Canberra Plan', *Pacific Philosophical Quarterly*, 92: 232–42.

Loewer, Barry (2007) 'Counterfactuals and the Second Law', in Huw Price and Richard Corry (eds), *Causation, Physics, and the Constitution of Reality. Russell's Republic Revisited* (Oxford: Clarendon Press): 293–326.

Loewer, Barry (2008) 'Why There Is Anything Except Physics', in Jakob Hohwy and Jesper Kallestrup (eds), *Being Reduced. New Essays on Reduction, Explanation, and Causation* (Oxford: Oxford University Press): 149–63.

Loewer, Barry (2009) 'Why Is There Anything Except Physics?', *Synthese*, 170: 217–33.

Lucas, Robert (1976) 'Econometric Policy Evaluation: A Critique', in K. Brunner and A. Meltzer (eds), *The Phillips Curve and Labor Markets*, Carnegie-Rochester

Conference Series on Public Policy, vol. 1, spring (Amsterdam: North-Holland): 161–8.

Mach, Ernst (1900) *Principien der Wärmelehre* (Leipzig: Barth).

Mach, Ernst (1980) *Erkenntnis und Irrtum* (Darmstadt: Wissenschaftliche Buchgesellschaft).

Mach, Ernst (1982) *Die Mechanik* (Darmstadt: Wissenschaftliche Buchgesellschaft).

Machamer, Peter (2004) 'Activities and Causation: The Metaphysics and Epistemology of Mechanisms', *International Studies in the Philosophy of Science*, 18: 27–39.

Machamer, Peter, Lindley Darden and Carl F. Craver (2000) 'Thinking about Mechanisms', *Philosophy of Science*, 67: 1–25.

Mackie, John Leslie (1965) 'Causes and Conditions', in Ernest Sosa and Michael Tooley (eds) (1993), *Causation* (Oxford: Oxford University Press) 33–55.

Mackie, John L. (1980) *The Cement of the Universe* (Oxford: Oxford University Press).

Marshall, Alfred (1890) *Principles of Economics*, 8th edn (Macmillan: London).

Mas-Colell, A., M. Whinston and J. Green (1995) *Microeconomic Theory* (Oxford: Oxford University Press).

Maudlin, Tim (2002) *Quantum Non-Locality and Relativity* (Oxford: Blackwell).

Maudlin, Tim (2004) 'Causation, Counterfactuals, and the Third Factor', in J. Collins, N. Hall and L. Paul, *Causation and Counterfactuals* (Cambridge, MA: MIT Press): 419–44.

Maudlin, Tim (2007) *The Metaphysics in Physics* (Oxford: Oxford University Press).

Mayntz, Renate (2002) *Akteure – Mechanismen – Modelle. Zur Theoriefähigkeit makro-sozialer Analysen* (Frankfurt: Campus).

Meek, C. and Clark Glymour (1994) 'Conditioning and Intervening', *British Journal for the Philosophy of Science*, 45: 1001–21.

Menzies, Peter (2007) 'Causation in Context', in Huw Price and Richard Corry (eds) (2007) *Causation, Physics, and the Constitution of Reality. Russell's Republic Revisited* (Oxford: Clarendon Press): 191–223.

Menzies, Peter and Huw Price (1993) 'Causation as a Secondary Quality', *British Journal of the Philosophy of Science*, 44: 187–203.

Mill, J.S. (1891) *A System of Logic* (New York: Harper).

Mill, J.S. (2008 [1836]) 'On the Definition and Method of Political Economy', in Daniel Hausman (ed.), *The Philosophy of Economics. An Anthology*, 3rd edn (New York: Cambridge University Press): 41–58.

Mitchell, Sandra (1997) 'Pragmatic Laws', *Philosophy of Science*, 64: 242–65.

Mitchell, Sandra (2000) 'Dimensions of Scientific Law', *Philosophy of Science*, 67: S468–S479.

Mitchell, Sandra (2002) 'Ceteris Paribus – An Inadequate Representation of Biological Contingency', *Erkenntnis*, 57: 329–50.

Mitchell, Sandra (2009) 'Complexity and Explanation in the Social Sciences', in C. Mantzavinos (ed.), *Philosophy of the Social Sciences. Philosophical Theory and Scientific Practice* (Cambridge: Cambridge University Press): 130–45.

Moore, Michael (2009) 'Introduction: The Nature of Singularist Theories of Causation', *The Monist*, 92: 3–22.

Morgan, Stephen and Christopher Winship (2007) *Counterfactuals and Causal Inference. Methods and Principles for Social Research* (Cambridge: Cambridge University Press).

Mumford, Stephen (1998) *Dispositions* (Oxford: Oxford University Press).

Mumford, Stephen (2004) *Laws in Nature* (London: Routledge).

Ney, Alyssa (2009) 'Physical Causation and Difference-Making', *British Journal for Philosophy of Science*, 60(4): 737–64.

North, Jill (2008) 'Two Views on Time Reversal', *Philosophy of Science*, 75: 201–23.

Norton, John (2007a) 'Causation as Folk Science', in Huw Price and Richard Corry (eds), *Causation, Physics and the Constitution of Reality* (Oxford: Oxford University Press): 11–44.

Norton, John (2007b) 'Do the Causal Principles of Modern Physical Contradict Causal Anti-Fundamentalism?', in Peter Machamer and Gereon Walters (eds), *Thinking about Causes* (Pittsburgh: Pittsburgh University Press): 222–34.

Norton, John (2009) 'Is There any Independent Principle of Causality in Physics?', *British Journal for Philosophy of Science*, 60: 475–86.

Okasha, Samir (2009) 'Causation in Biology', in Helen Beebee, Christopher Hitchcock and Peter Menzies (eds), *The Oxford Handbook of Causation* (Oxford: Oxford University Press): 707–25.

Pearl, Judea (1988) *Probabilistic Reasoning in Intelligent Systems* (San Mateo, CA: Morgan Kaufmann).

Pearl, Judea (1994) 'From Adams' Conditionals to Default Expressions, Causal Conditionals, and Counterfactuals', in Ellery Eells and Brian Skyrms (eds), *Probability and Conditionals. Belief Revision and Rational Decision* (Cambridge, UK: Cambridge University Press): 47–74.

Pearl, Judea (2000) *Causality: Models, Reasoning and Inference* (Cambridge: Cambridge University Press).

Pearl, Judea and D. Galles (1996) 'Axioms of Causal Relevance', UCLA Cognitive Systems Laboratory, Technical Report (R-240-S), January.

Pearl, Judea and D. Galles (1998) 'An Axiomatic Characterization of Causal Counterfactuals', *Foundations of Science*, 3: 151–82.

Popper, Karl (1934) *Logik der Forschung* (Tübingen: Mohr Siebeck).

Pietroski, Paul and Georges Rey (1995) 'When Other Things Aren't Equal: Saving Ceteris Paribus Laws from Vacuity', *British Journal for the Philosophy of Science*, 46: 81–110.

Price, Huw (2007) 'Causal Perpectivalism', in Huw Price and Richard Corry (eds), *Causation, Physics, and the Constitution of Reality. Russell's Republic Revisited* (Oxford: Clarendon Press): 250–92.

Price, Huw and Richard Corry (eds) (2007) *Causation, Physics, and the Constitution of Reality. Russell's Republic Revisited* (Oxford: Clarendon Press).

Psillos, Stathis (2002) *Causation and Explanation* (Chesham: Acumen).

Psillos, Stathis (2004) 'A Glimpse of the Secret Connexion: Harmonising Mechanisms with Counterfactuals', *Perspectives on Science*, 12: 288–319.

Psillos, Stathis (2007) 'Causal Explanation and Manipulation', in Johannes Person and Petri Ylikoski (eds), *Rethinking Explanation*, Boston Studies in the Philosophy of Science, vol. 252, Springer: 97–112.

Psillos, Stathis (2009) 'Regularity Theories', in H. Beebee, C. Hitchcock and P. Menzies (eds), *Oxford Handbook of Causation* (New York: Oxford University Press) 133–57.

Putnam, Hilary (1975) 'Philosophy of Logic', repr. in *Mathematics Matter and Method: Philosophical Papers, Volume 1*, 2nd edn (Cambridge: Cambridge University Press): 323–57.

Quine, O.W.V. (1960) *Word and Object* (Cambridge, MA: MIT Press).

Ramsey, Frank P. (1929) 'General Propositions and Causality', in D.H. Mellor (ed.), *F. P. Ramsey: Philosophical Papers*: 237–55.

Redhead, Michael (1990) 'Explanation', in D. Knowles, (ed.), *Explanation and Its Limits* (Cambridge: Cambridge University Press): 135–54.

Reichenbach, Hans (1956) *The Direction of Time* (Mineola, NY: Dover Publications).

Reiss, Julian (2008) *Error in Economics* (London: Routledge).

Reutlinger, Alexander, Andreas Hüttemann and Gerhard Schurz (2011) 'Ceteris Paribus Laws', in Edward N. Zalta (ed.), *Stanford Encyclopedia of Philosophy* (spring 2011 edn), available at http://plato.stanford.edu/archives/spr2011/entries/ceteris-paribus/.

Reutlinger, A. (2013) 'Can Interventionists be Neo-Russellians?' unpublished MS, under review.

Roberts, John (2004) 'There are No Laws in the Social Sciences', in Christopher Hitchcock (ed.), *Contemporary Debates in the Philosophy of Science* (Oxford: Blackwell): 168–85.

Rosenberg, Alexander (2001) 'How Is Biological Explanation Possible?', *British Journal for Philosophy of Science*, 52: 735–60.

Rosenberg, Alexander (2009) 'If economics is a science, what kind of science is it?', in H. Kincaid and D. Ross (eds), *The Oxford Handbook of Economics* (Oxford: Oxford University Press): 55–67.

Rosenberg, Alexander and Daniel W. McShea (2008) *Philosophy of Biology: A Contemporary Introduction* (London: Routledge).

Ross, Don and David Spurrett (2007) 'Notions of Cause. Russell's Thesis Revisited', *British Journal for Philosophy of Science*, 58(1): 45–76.

Russell, Bertrand (1912/13) 'On the Notion of Cause', *Proceedings of the Aristotelian Society*, 13: 1–26.

Russo, Federica (2009) *Causality and Causal Modelling in the Social Sciences. Measuring Variations* (New York: Springer).

Ryle, Gilbert (1949) *The Concept of Mind* (Chicago University of Chicago Press).

Salmon, Wesley (1984) *Scientific Explanation and the Causal Structure of the World* (Princeton: Princeton University Press).

Salmon, Wesley (1989) 'Four Decades of Explanation', in P. Kitcher and W. Salmon (eds), *Scientific Explanation, Minnesota Studies in the Philosophy of Science XVIII* (Minneapolis: University of Minnesota Press): 3–219.

Schaffer, Jonathan (2004a) 'Trumping Preemption', in J. Collins, N. Hall and L. Paul, *Causation and Counterfactuals* (Cambridge, MA: MIT Press): 59–74.

Schaffer, Jonathan (2004b) 'Counterfactuals, Causal Independence, and Conceptual Circularity', *Analysis*, 64(4): 299–309.

Scheibe, Erhard (2006) *Die Philosophie der Physiker* (München: Beck).

Scheines, Richard (1997) 'An Introduction to Causal Inference', in Vaughan McKim (ed.), *Causality in Crisis* (Notre Dame: University of Notre Dame Press): 185–99.

Schrenk, Markus (2007) *The Metaphysics of Ceteris Paribus Laws* (Frankfurt: Ontos).

Schurz Gerhard (1997) 'Probabilistic Default Reasoning Based on Relevance and Irrelevance Assumptions', in D. Gabbay, R. Kruse, A. Nonnengart and H. J. Ohlbach (eds), *Qualitative and Quantitative Practical Reasoning* (LNAI 1244) (Berlin: Springer): 536–53.

Schurz, Gerhard (1998) 'Probabilistic Semantics for Delgrande's Conditional Logic and a Counterexample to his Default Logic', *Artificial Intelligence*, 102(1): 81–95.

Schurz, Gerhard (2001) 'Pietroski and Rey on Ceteris Paribus Laws', *British Journal for Philosophy of Science*, 52: 359–70.

Schurz, Gerhard (2002) 'Ceteris Paribus Laws: Classification and Deconstruction', *Erkenntnis*, 57: 351–72.

Schurz, Gerhard (2006) *Einführung in die Wissenschaftstheorie* (Darmstadt: Wissenschaftliche Buchgesellschaft).

Shapiro, L. (forthcoming) 'Lessons from Causal Exclusion', *Philosophy and Phenomenological Research*.

Shapiro, L. and Sober, E. (2007) 'Epiphenomenalism. The Dos and Don'ts', in G. Wolters and P. Machamer (eds), *Thinking about Causes: From Greek Philosophy to Modern Physics* (Pittsburgh: University of Pittsburgh Press): 235–64.

Sider, Theodore (2010) *Logic For Philosophy* (Oxford: Oxford University Press).

Sinai, Yakov G. (1992) *Probability Theory. An Introductory Course* (Berlin/New York: Springer).

Skyrms, Brian (1980a) 'The Prior Propensity Account of Subjunctive Conditionals', in W. Harper, G. Pearse and Robert Stalnaker (eds), *IFS: Conditionals, Belief, Decision, Chance and Time* (Dordrecht: Reidel): 259–65.

Skyrms, Brian (1980b) *Causal Necessity* (New Haven: Yale University Press).

Skyrms, Brian (1984) *Pragmatism and Empiricism* (New Haven: Yale University Press).

Skyrms, Brian (1994) 'Adams Conditionals', in Ellery Eells and Brian Skyrms (eds), *Probability and Conditionals. Belief Revision and Rational Decision* (Cambridge, UK: Cambridge University Press): 13–26.

Sober, Elliott (1997) 'Two Outbreaks of Lawlessness in Recent Philosophy of Biology', *Philosophy of Science*, 64: 432–44.

Spirtes, Peter, Clark Glymour and Richard Scheines (2000) *Causation, Prediction and Search* (New York: Springer).

Spohn, Wolfgang (forthcoming) *Ranking Theory. A Tool for Epistemology* (Oxford: Oxford University Press).

Stalnaker, Robert (1968) 'A Theory of Conditionals', in *IFS: Conditionals, Belief, Decision, Chance and Time* (Dordrecht: Reidel): 41–55.

Steel, Daniel (2005) 'Indeterminism and the Causal Markov Condition', *British Journal for Philosophy of Science*, 56: 3–26.

Steel, Daniel (2006) 'Comment on Hausman and Woodward on the Causal Markov Condition', *British Journal for Philosophy of Science*, 57: 219–31.

Steel, Daniel (2007) *Across the Boundaries. Extrapolation in Biology and Social Science* (New York: Oxford University Press).

Strawson, Peter (1992) *Analysis and Metaphysics* (Oxford: Oxford University Press).

Strevens, Michael (2003) *Bigger Than Chaos. Understanding Complexity through Probability* (Cambridge, MA: Harvard University Press).

Strevens, Michael (2007) 'Review of Woodward, Making Things Happen', *Philosophy and Phenomenological Research*, 74(1): 233–49.

Strevens, Michael (2008) 'Comments on Woodward, Making Things Happen', *Philosophy and Phenomenological Research*, 77(1): 171–92.

Strevens, Michael (2009) *Depth. An Account of Scientific Explanation* (Cambridge, MA: Harvard University Press).

Strevens, Michael (2010) 'Ceteris Paribus Hedges: Causal Voodoo That Works', available at http://www.strevens.org/research/lawmech/CPMechBrev.pdf.

Suppes, Patrick (1957) *Introduction to Logic* (Princeton: D. Van Nostrand).

Suppes, Patrick (1970) *A Probabilistic Theory of Causality* (Amsterdam: North-Holland).

Tooley, Michael (1977) 'The Nature of Laws', *Canadian Journal of Philosophy*, 7: 667–98.

Tooley, Michael (2009) 'Causes, Laws, and Ontology', in Helen Beebee, Christopher Hitchcock and Peter Menzies (eds), *The Oxford Handbook of Causation* (Oxford: Oxford University Press): 368–86.

Wasserman, Ryan (2006) 'The Future Similarity Objection Revisited', *Synthese*, 150: 57–67.

Waters, Ken (2007) 'Causes that Make a Difference', *Journal of Philosophy*, 104(11): 551–79.

Weber, Marcel (2008) 'Causes without Mechanisms: Experimental Regularities, Physical Laws, and Neuroscientific Explanation', *Philosophy of Science*, 75: 995–1007.

Wien, W. (1915) 'Ziele und Methoden der theoretischen Physik', *Jahrbuch der Radioaktivität und Elektronik* 12: 241–59.

Williamson, Jon (2005) *Bayesian Nets and Causality. Philosophical and Computational Foundations* (Oxford: Oxford University Press).

Woodward, James (1992) 'Realism about Laws', *Erkenntnis*, 36: 181–218.

Woodward, James (2000) 'Explanation and Invariance in the Special Sciences', *British Journal for the Philosophy of Science*, 51: 197–254.

Woodward, James (2002) 'There is No Such Thing as a Ceteris Paribus Law', *Erkenntnis*, 57: 303–28.

Woodard, James (2003) *Making Things Happen. A Theory of Causal Explanation* (Oxford: Oxford University Press).

Woodward, James (2004) 'Counterfactuals and Causal Explanation', *International Studies in the Philosophy of Science*, 18: 41–72.

Woodward, James (2007) 'Causation with a Human Face', in Huw Price and Richard Corry (eds), *Causation, Physics, and the Constitution of Reality. Russell's Republic Revisited* (Oxford: Clarendon Press): 66–105.

Woodward, James (2008) 'Response to Strevens', *Philosophy and Phenomenological Research*, 77(1): 193–212.

Woodward, James (2009) 'Scientific Explanation', Edward N. Zalta (ed.), *Stanford Encyclopedia of Philosophy*, spring 2010 edn, available at http://plato.stanford.edu/archives/spr2010/entries/scientific-explanation/.

Woodward, James and Christopher Hitchcock (2003) 'Explanatory Generalizations, Part I: A Counterfactual Account', *Noûs*, 37(1): 1–24.

Wright, Georg Henrik von (1971) *Explanation and Understanding* (Cornell University Press: Ithaca).

Wright, Georg H. von (1974) *Causality and Determinism* (New York/London: Columbia University Press).

# Index

Printed in the United States
by Baker & Taylor Publisher Services